AF709871

BLAST OFF to 4th Grade Math

3rd to 4th Grade Transition Math Workbook

This book belongs to:

____

- - - - - - - - - - - - - - - -

____

GET THE ANSWER KEY

The answer key for this book is available as a free PDF download.

SCAN HERE

Copyright © 2023 EDventure Learning LLC

All rights reserved. This book or any portion thereof may not be reproduced or used in any manner whatsoever without the express written permission of the publisher.

This product is not sponsored or endorsed by the Common Core State Standards Initiative of the National Governors Association Center for Best Practices and the Council of Chief State School Officers.

ISBN: 978-1-64824-040-9
EDventure Learning LLC
5601 State Route 31 #1296
Clay, NY 13039

www.edventurelearning.com

Email us at hello@edventurelearning.com
Follow us on social media @edventurelearning

3 - 4

This book is part of the
Bridge Builder Learning Series,
designed to help bridge the gap between grade levels. The content in this book is designed to review math from 3rd grade and set students up for success in 4th grade.

CONTENTS

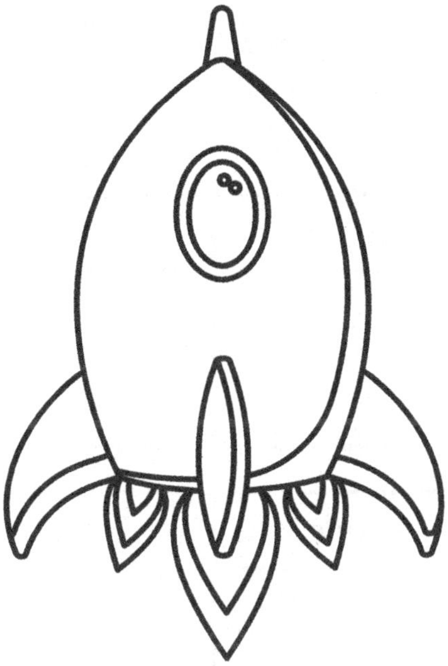

Numbers and Place Value

6 Place Value

For each number, write the digits in the correct columns in the place value chart.

57,431

Ten thousands	Thousands	Hundreds	Tens	Ones
5	7	4	3	1

82,047

Ten thousands	Thousands	Hundreds	Tens	Ones

10,678

Ten thousands	Thousands	Hundreds	Tens	Ones

77,810

Ten thousands	Thousands	Hundreds	Tens	Ones

6,546

Ten thousands	Thousands	Hundreds	Tens	Ones

31,005

Ten thousands	Thousands	Hundreds	Tens	Ones

Write each number in expanded form.

1. 25,987 = _____ + _____ + _____ + _____ + _____

2. 16,721 = _____ + _____ + _____ + _____ + _____

3. 32,945 = _____ + _____ + _____ + _____ + _____

4. 45,263 = _____ + _____ + _____ + _____ + _____

5. 19,819 = _____ + _____ + _____ + _____ + _____

Change each expanded form into a number in standard form.

6. 30,000 + 1,000 + 700 + 80 + 9 = __________

7. 80,000 + 2,000 + 100 + 70 + 2 = __________

8. 5,000 + 900 + 30 + 8 = __________

9. 40,000 + 500 + 90 + 3 = __________

10. 50,000 + 4,000 + 600 + 10 + 5 = __________

8 Place Value

Write each number in standard form.

1. One thousand three hundred twenty-six ____________

2. Six thousand two hundred ninety-seven ____________

3. Nine hundred eighteen ____________

4. Fifty-two thousand four hundred eleven ____________

5. Eighty-one thousand thirty-five ____________

Write out the name of each number.

6. 8,457 ______________________________

7. 21,860 ______________________________

8. 13,145 ______________________________

9. 5,092 ______________________________

10. 48,917 ______________________________

Place Value

Complete the diagram about each number.

8,271

NAME

PLACE VALUES

_____ thousands _____ tens

_____ hundreds _____ ones

EXPANDED FORM

_____ + _____ + _____ + _____

5,148

NAME

PLACE VALUES

_____ thousands _____ tens

_____ hundreds _____ ones

EXPANDED FORM

_____ + _____ + _____ + _____

1,693

NAME

PLACE VALUES

_____ thousands _____ tens

_____ hundreds _____ ones

EXPANDED FORM

_____ + _____ + _____ + _____

10 Place Value

Write the value of the underlined digit in each number.

(1) 56,0**9**1 ____________

(2) 34,08**7** ____________

(3) 7**8**,362 ____________

(4) 49,5**1**3 ____________

(5) 20,91**5** ____________

(6) **6**8,761 ____________

(7) 12,**4**56 ____________

(8) 81,5**6**0 ____________

(9) 91,**2**38 ____________

(10) 9**3**,108 ____________

Circle the number in each pair in which the digit **5** has a **greater** value.

(11) 52,487 35,290

(12) 15,921 18,579

(13) 12,345 87,654

(14) 10,592 45,012

(15) 98,572 65,081

Place Value

Fill in the blanks with the correct numbers.

1. 5 tens = ______ ones

2. 3 hundreds = ______ tens

3. 2 thousands = ______ hundreds

4. 8 ten thousands = ______ thousands

5. 9 hundreds = ______ ones

6. 500 ones = ______ tens

7. 20 hundreds = ______ thousands

8. 1 thousand = ______ tens

9. 30 thousands = ______ ten thousands

10. 7 thousands = ______ ones

12 Place Value

Fill the in correct bubble to answer each question.

What does the 1 represent in the number 51,247?

- ○ 10,000
- ○ 1,000
- ○ 100

What does the 9 represent in the number 38,492?

- ○ 9 ones
- ○ 9 tens
- ○ 9 hundreds

What does the 8 represent in the number 19,845?

- ○ 8,000
- ○ 80
- ○ 800

Which digit in is the ten thousands place in the number 13,974?

- ○ 1
- ○ 3
- ○ 7

Which digit is in the hundreds place in the number 75,209?

- ○ 5
- ○ 7
- ○ 2

Place Value

Write the four-digit number that answers each riddle.

My tens and ones digits add up to my hundreds digit. My thousands digit is one less than my tens digit. My ones digit is 5 and my thousands digit is 1. What number am I?	
My thousands digit and ones digit are the same. The value of my hundreds digit is 200. My last two digits are 46. What number am I?	
My thousands digit is 5 and my tens digit is 1. My tens and ones digits add up to my thousands digit. My hundreds digit is one less than my tens digit. What number am I?	
All of my digits add up to 9. My hundreds digit and tens digit are the same. My thousands digit is 4 and my ones digit is 1. What number am I?	
My largest digit is in the ones place. I am an even number. Three of my digits are 7. What number am I?	

14 Ordering and Comparing

Write each set of numbers in the train cars in order from least to greatest.

(EX) **5,120** **5,801** **4,910** **4,936**

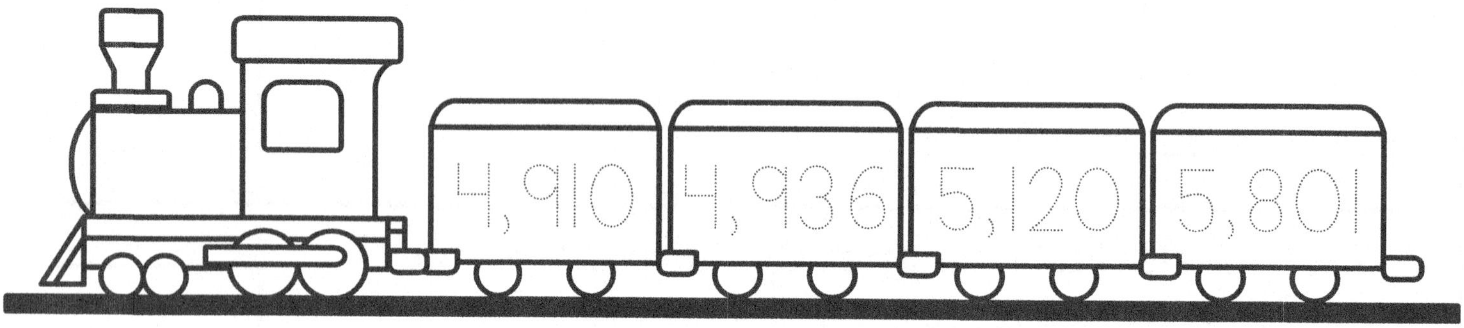

(1) **18,026** **20,451** **17,389** **9,678**

(2) **219** **1,057** **689** **2,176**

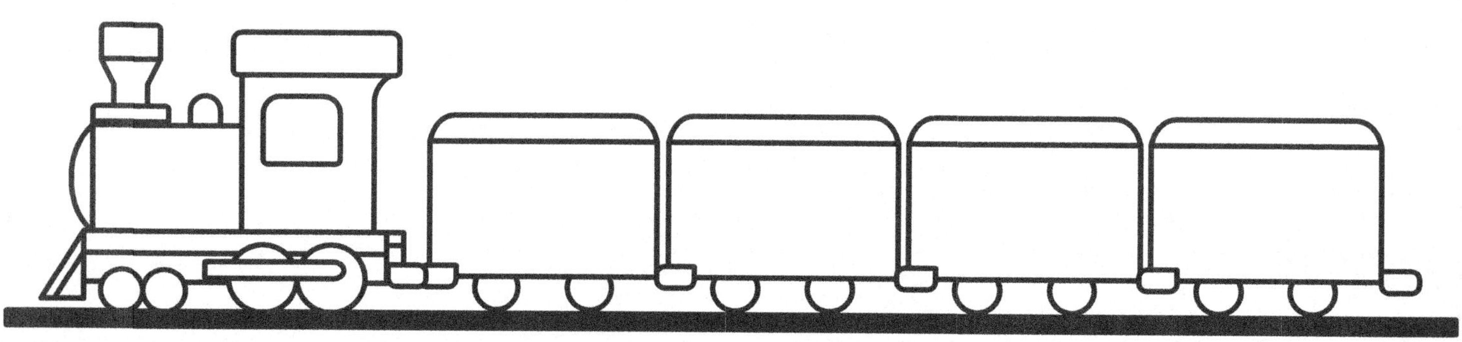

3. **65,123** **56,987** **59,120** **60,791**

4. **9,052** **8,173** **10,328** **9,287**

5. **17,140** **17,410** **17,389** **17,399**

Ordering and Comparing

Compare each pair of numbers using one of these symbols:

> greater than < less than = equal to

1,023	>	1,017	7,981		7,962
5,170		4,910	523		524
7,120		7,120	2,750		2,573
2,718		2,728	21,417		19,456
12,920		13,081	952		952
11,259		11,257	87,461		78,461
51,942		52,318	55,555		7,777

Answer each word problem.

Lucas scored lower on the math test than Miguel. David scored higher on the math test than Miguel. Cameron scored lower on the test than Lucas. Whose score was the highest?	________________
Mount Whitney is 14,505 feet tall. Mount Fairweather is 15,325 feet tall. Mount Rainier is 14,411 feet tall. Mount Hubbard is 14,951 feet tall. List the names of the mountains from tallest to shortest.	Tallest: ________________ ________________ ________________ Shortest: ________________
Ava, Brianna, Chelsea, and Dina ran a race. Dina finished before Ava. Chelsea came in 2nd place. Ava finished before Brianna. In what order did the runners finish the race?	1st place: ________________ 2nd place: ________________ 3rd place: ________________ 4th place: ________________

18 Rounding

Use the number line to help you round each number to the nearest ten.

① 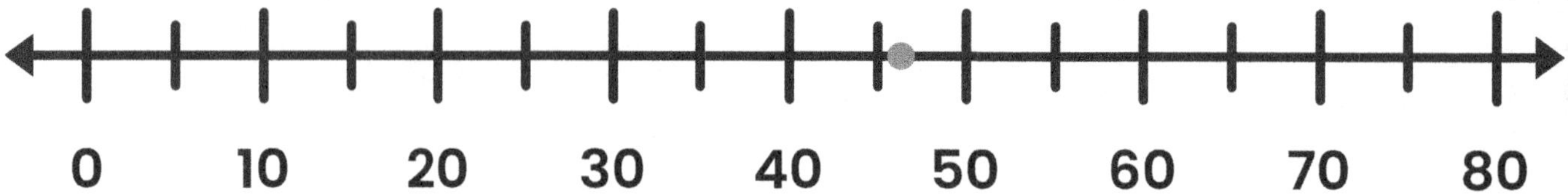

46 rounded to the nearest ten is ______

② 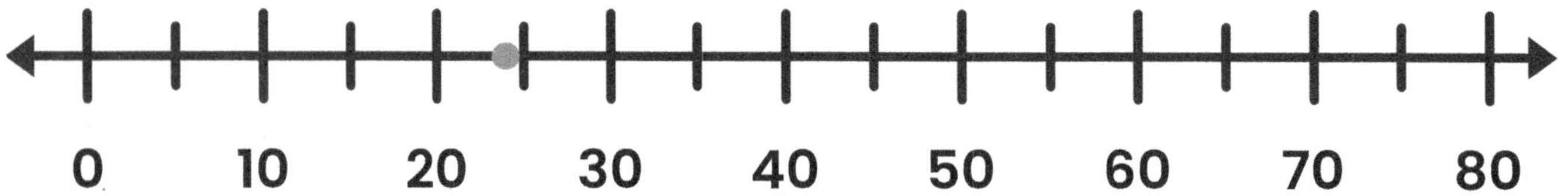

24 rounded to the nearest ten is ______

③ 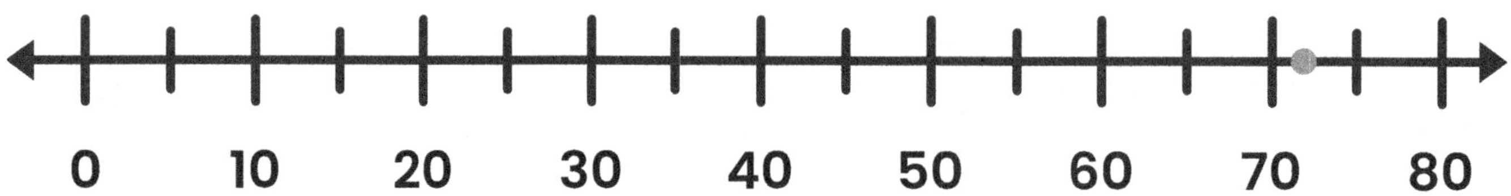

72 rounded to the nearest ten is ______

④ 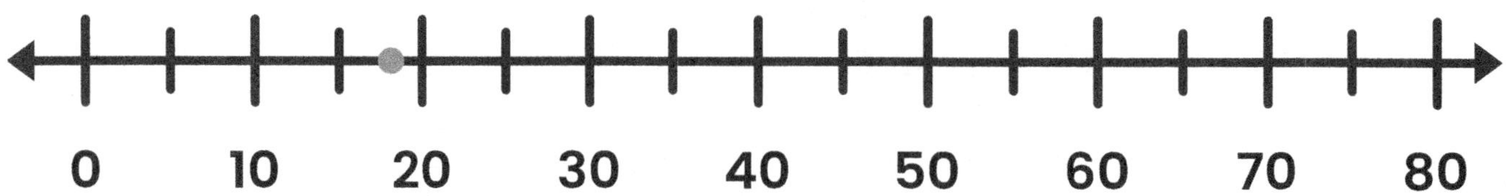

18 rounded to the nearest ten is ______

Rounding

Use the number line to help you round each number to the nearest hundred.

1 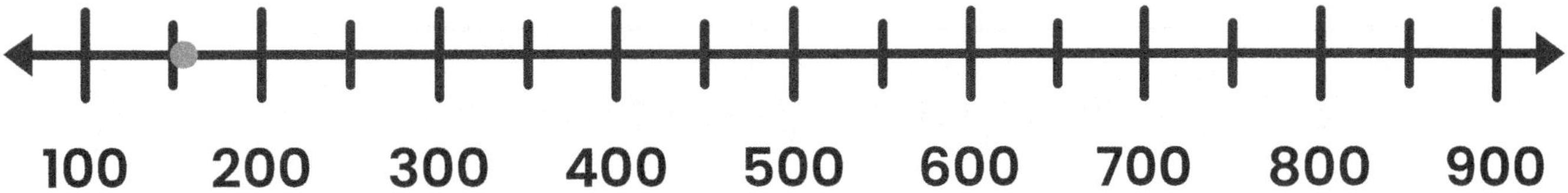

154 rounded to the nearest hundred is ______

2 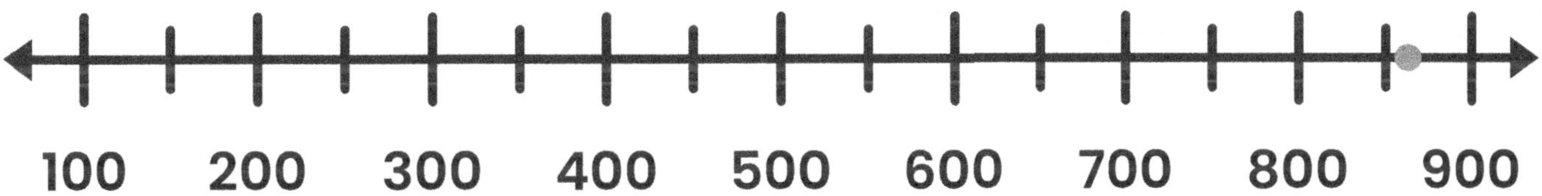

867 rounded to the nearest hundred is ______

3 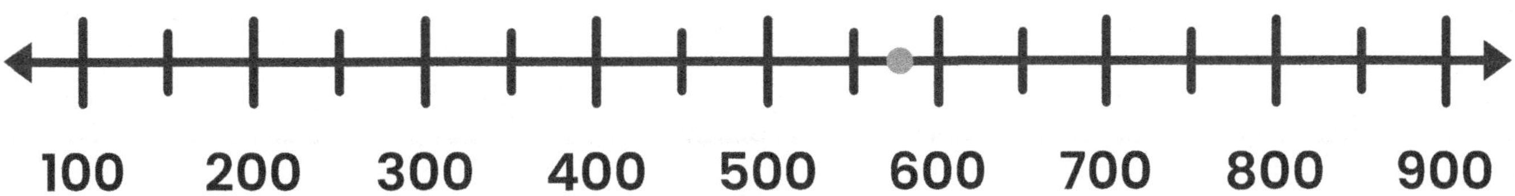

579 rounded to the nearest hundred is ______

4 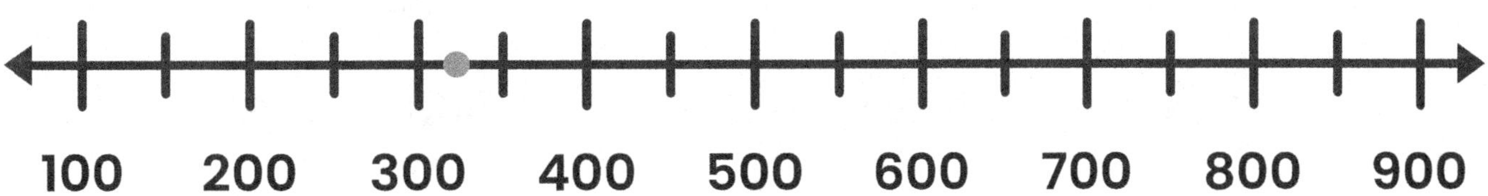

320 rounded to the nearest hundred is ______

20 Rounding

Fill the in correct bubble to answer each question.

What is 287 rounded to the nearest ten?

- ○ 280
- ○ 290
- ○ 300

What is 1,451 rounded to the nearest hundred?

- ○ 1,400
- ○ 1,450
- ○ 1,500

Which of these numbers would round to 480 as the nearest ten?

- ○ 486
- ○ 478
- ○ 474

Which of these numbers would round to 900 as the nearest ten?

- ○ 897
- ○ 909
- ○ 923

Which of these numbers would round to 500 as the nearest hundred?

- ○ 412
- ○ 549
- ○ 570

Rounding

Round each number to the nearest ten and the nearest hundred.

	Nearest Ten	Nearest Hundred
605	610	600

	Nearest Ten	Nearest Hundred
2,319		

	Nearest Ten	Nearest Hundred
987		

	Nearest Ten	Nearest Hundred
723		

	Nearest Ten	Nearest Hundred
1,481		

	Nearest Ten	Nearest Hundred
564		

	Nearest Ten	Nearest Hundred
175		

	Nearest Ten	Nearest Hundred
212		

	Nearest Ten	Nearest Hundred
7,916		

	Nearest Ten	Nearest Hundred
338		

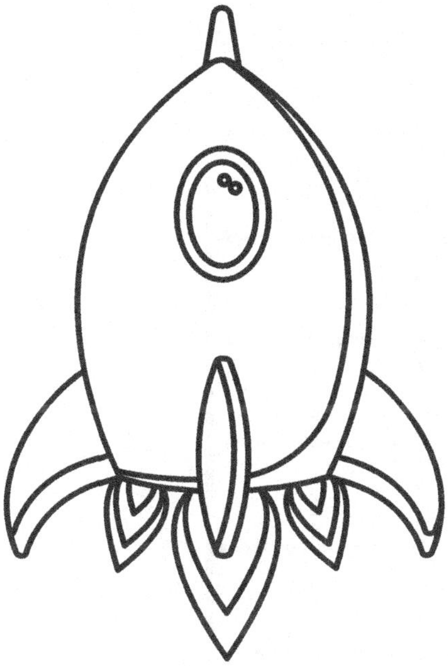

Operations

Addition and Subtraction

Complete the addition table.

+	1	2	3	4	5	6	7	8	9
1									
2									
3									
4									
5									
6									
7									
8									
9									

Which of these will always give you an even sum?
(select all that apply)

- ○ even + even
- ○ even + odd
- ○ odd + odd

Which of these will always give you an odd sum?
(select all that apply)

- ○ even + even
- ○ even + odd
- ○ odd + odd

Fill in the correct bubble to answer each question.

Which of these expressions is equal to 6 + 14 ?

○ 14 - 6 ○ 14 + 6 ○ 20 - 6

Which of these expressions is NOT equal to 15 + 22 ?

○ 30 + 7 ○ 20 + 17 ○ 35 + 7

Which of these expressions is NOT equal to 42 + 17 + 6 ?

○ 42 + 23 ○ 60 + 4 ○ 59 + 6

Which of these expressions is equal to 83 - 38 ?

○ 80 - 35 ○ 45 + 38 ○ 85 - 35

Which of these expressions is NOT equal to 25 + 37 ?

○ 20 + 30 + 12 ○ 25 + 30 + 7 ○ 30 + 20 + 7

Addition and Subtraction

For each set of three numbers given, write four related addition and subtraction facts.

①

10	25	15

____ + ____ = ____

____ + ____ = ____

____ - ____ = ____

____ - ____ = ____

②

37	13	24

____ + ____ = ____

____ + ____ = ____

____ - ____ = ____

____ - ____ = ____

Find each sum or difference. Regroup if needed.

45	52	24	76	88	62
+ 33	+ 19	+ 61	+ 87	+ 10	+ 18
78					

29	76	43	82	70	65
− 17	− 43	− 15	− 41	− 32	− 28

21	74	57	35	18	78
+ 38	− 23	+ 14	− 12	+ 22	− 56

80	45	51	64	94	46
− 60	+ 52	− 18	+ 33	− 52	+ 49

74	67	92	28	85	39
+ 14	+ 23	− 46	+ 37	− 53	− 18

Addition and Subtraction

Estimate each sum or difference by rounding to the nearest hundred.

EX

		Estimate
422	→	400
+ 287	→	+ 300
		700

1

		Estimate
593	→	
+ 634	→	+ ____

2

		Estimate
805	→	
− 142	→	− ____

3

		Estimate
375	→	
+ 726	→	+ ____

4

		Estimate
910	→	
− 133	→	− ____

5

		Estimate
689	→	
+ 92	→	+ ____

6

		Estimate
876	→	
− 204	→	− ____

7

		Estimate
359	→	
+ 442	→	+ ____

Estimate. Then find each sum or difference.

(EX) Estimate: 1,000

$$\begin{array}{r} 341 \\ +\ 672 \\ \hline 1{,}013 \end{array} \qquad \begin{array}{r} 300 \\ +\ 700 \\ \hline 1{,}000 \end{array}$$

(1) Estimate: ________

$$\begin{array}{r} 109 \\ +\ 829 \\ \hline \end{array}$$

(2) Estimate: ________

$$\begin{array}{r} 278 \\ +\ 722 \\ \hline \end{array}$$

(3) Estimate: ________

$$\begin{array}{r} 492 \\ +\ 384 \\ \hline \end{array}$$

(4) Estimate: ________

$$\begin{array}{r} 926 \\ -\ 504 \\ \hline \end{array}$$

(5) Estimate: ________

$$\begin{array}{r} 764 \\ -\ 259 \\ \hline \end{array}$$

(6) Estimate: ________

$$\begin{array}{r} 619 \\ -\ 386 \\ \hline \end{array}$$

(7) Estimate: ________

$$\begin{array}{r} 838 \\ -\ 427 \\ \hline \end{array}$$

30 Addition and Subtraction

Fill in the missing number to make each equation true.

(EX) 25 = 13 + 12

1. 302 + ______ = 320

2. 697 - ______ = 462

3. 160 = ______ + 40

4. ______ - 287 = 512

5. 427 + ______ = 603

6. 978 - ______ = 517

7. 481 = ______ + 270

8. ______ - 85 = 106

Fill in the correct bubble to answer each question.

What is the sum of 319 and 562?

- ○ 871
- ○ 879
- ○ 881

Which of these does NOT equal 427?

- ○ 210 + 217
- ○ 311 + 116
- ○ 370 + 47

What is the difference between 645 and 268?

- ○ 913
- ○ 377
- ○ 423

Which of these does NOT equal 361?

- ○ 845 - 524
- ○ 484 - 123
- ○ 812 - 451

Which of these expressions has the greatest value?

- ○ 615 - 157
- ○ 234 + 376
- ○ 781 - 460

Addition and Subtraction

Add or subtract to solve each word problem.

The fourth grade teachers at a school are Mrs. Green, Mr. Miller, Ms. Jones, and Mrs. Smith. There are 25 students in Mrs. Green's class, 21 in Mr. Miller's class, 23 in Ms. Jones' class, and 20 in Mrs. Smith's class. How many fourth graders are there all together?

________ fourth graders

For a reading challenge, Jessica has read 326 pages so far. Nico has read 298 pages. How many more pages has Jessica read than Nico?

________ pages

A football team scores 14 points in the first quarter of a game, no points in the second quarter, 7 points in the third quarter, and 6 points in the fourth quarter. What is their final score in the game?

________ points

The first book in a series has 276 pages, the second book has 312 pages, and the third has 352 pages. How many pages are there in the entire book series?

_______ pages

Morgan and Danielle live in different cities and are meeting up in Atlanta. Morgan lives 457 miles from Atlanta, and Danielle lives 384 miles from Atlanta. How much farther does Morgan have to travel than Danielle?

_______ miles

Fiona and Gretchen are selling cookies for a fundraiser. Fiona has sold 145 boxes of cookies, and Gretchen has sold 219 boxes of cookies. How many more boxes has Gretchen sold than Fiona?

_______ boxes of cookies

34 Multiplication

Fill in the blanks to write an addition equation and a multiplication equation to describe each picture.

(EX)

__6__ + __6__ + __6__ + __6__ = __24__

__4__ × __6__ = __24__

(1)

______ + ______ + ______ + ______ = ______

______ × ______ = ______

(2)

______ + ______ + ______ = ______

______ × ______ = ______

3 

______ + ______ + ______ = ______

______ × ______ = ______

4 

______ + ______ + ______ + ______ + ______ = ______

______ × ______ = ______

5 

______ + ______ = ______

______ × ______ = ______

36 Multiplication

Complete the multiplication equation to represent each set of tens.

EX

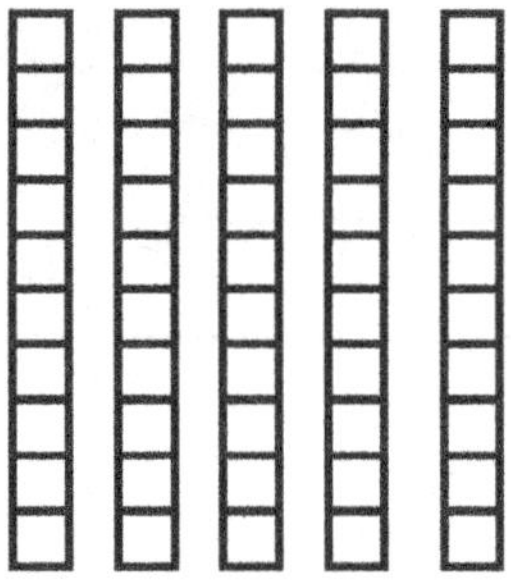

5 × 10 = 50

1

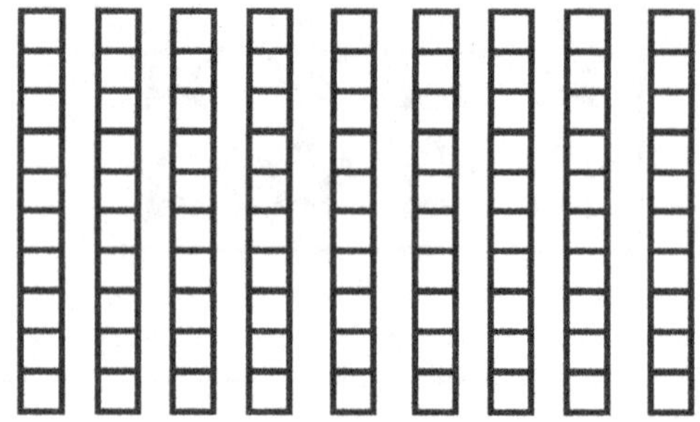

_____ × 10 = _____

2

_____ × 10 = _____

3

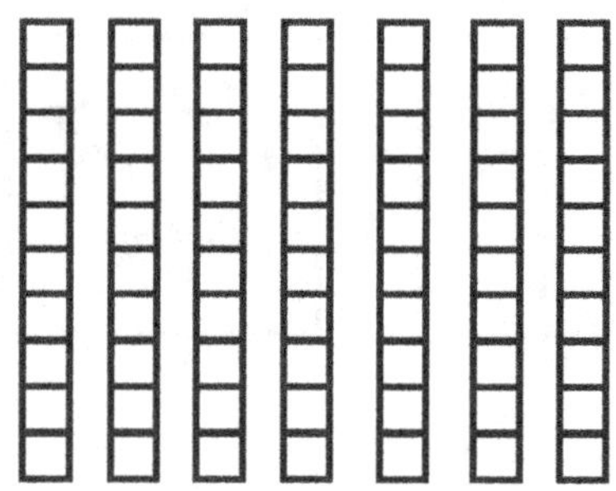

_____ × 10 = _____

4

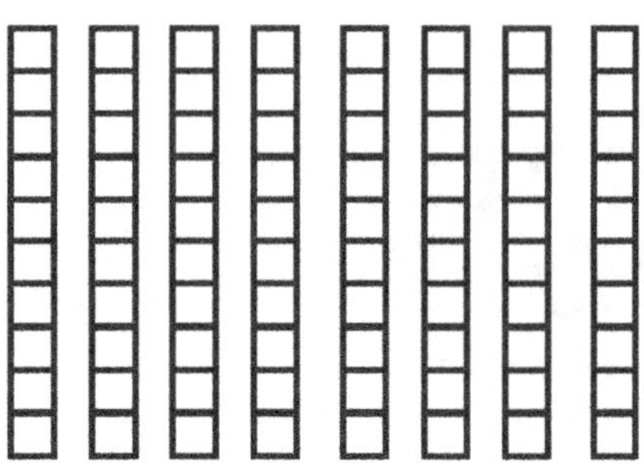

_____ × 10 = _____

5

_____ × 10 = _____

Multiplication

Select the multiplication expression that best represents each picture.

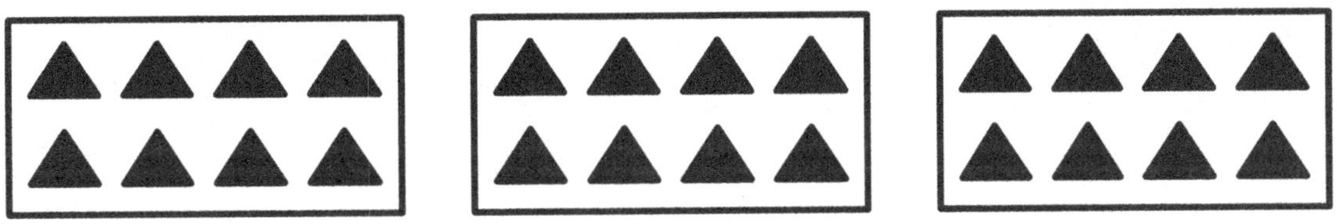

○ 3 × 8 ○ 8 × 2 ○ 4 × 2

○ 3 × 6 ○ 5 × 3 ○ 3 × 3

○ 2 × 2 ○ 4 × 2 ○ 4 × 4

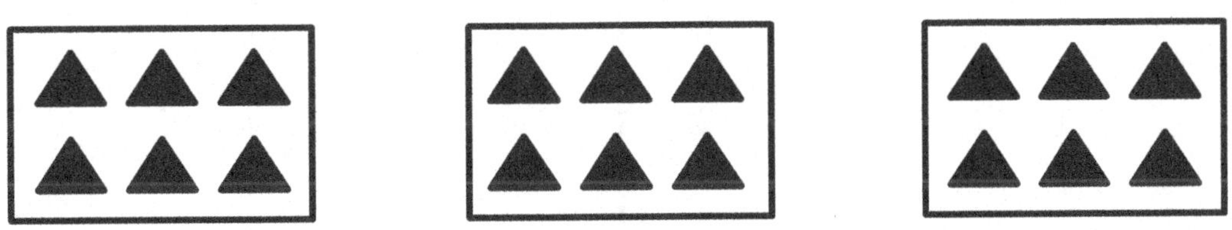

○ 3 × 6 ○ 6 × 2 ○ 3 × 7

○ 2 × 6 ○ 7 × 7 ○ 7 × 2

38 Multiplication

Draw a picture to represent each multiplication expression, then solve.

$4 \times 10 =$ _____	$2 \times 5 =$ _____
$3 \times 7 =$ _____	$3 \times 2 =$ _____
$5 \times 6 =$ _____	$7 \times 2 =$ _____
$8 \times 3 =$ _____	$4 \times 9 =$ _____

Fill in the correct bubbles to answer each question. Some questions have more than one correct answer.

Which of these expressions is equal to 4×3 ?

○ 3×4 ○ $4 + 4 + 4$ ○ $3 + 3 + 3$

Which of these expressions is equal to 2×8 ?

○ $8 + 2$ ○ 8×2 ○ $8 + 8$

Which of these is equal to 5×3 ?

○ 2 ○ 8 ○ 15

Which of these is equal to 2×6 ?

○ 8 ○ 12 ○ 18

Which of these is equal to 4×2 ?

○ 2 ○ 6 ○ 8

40 Multiplication

Write a multiplication equation to describe each array.

(EX) 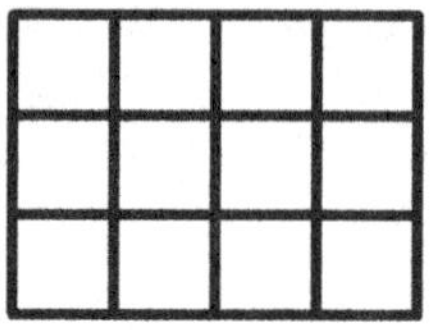

3 × 4 = 12

(1) 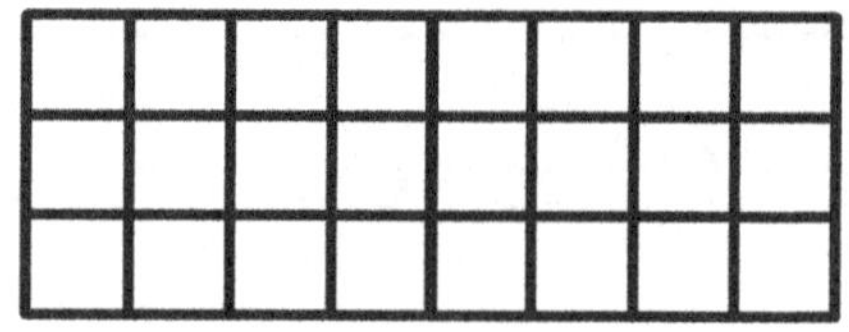

______ × ______ = ______

(2) 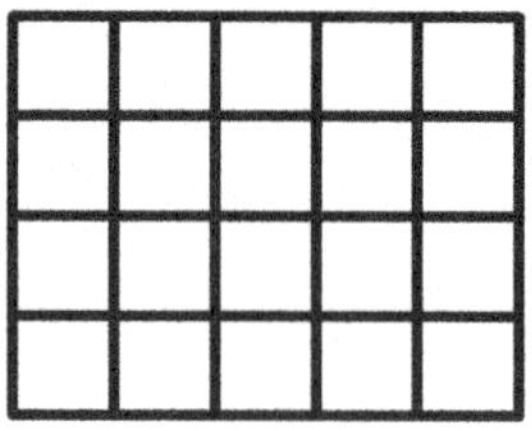

______ × ______ = ______

(3) 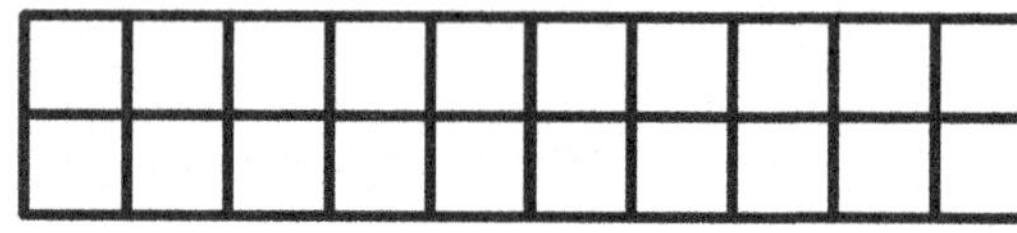

______ × ______ = ______

(4) 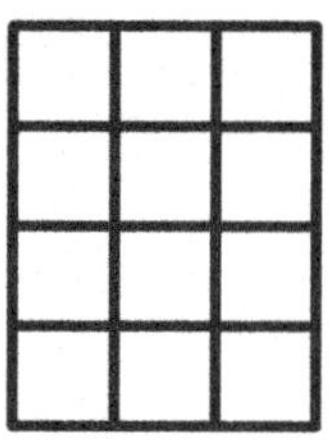

______ × ______ = ______

(5) 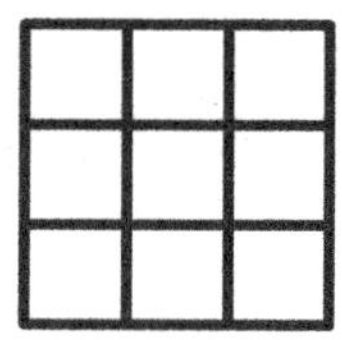

______ × ______ = ______

Draw an array to show each multiplication equation, then solve.

(6) 2 × 5 = ______

(7) 4 × 7 = ______

(8) 3 × 9 = ______

Draw an array to help you solve each word problem.

A sheet of stickers has 8 rows with 6 stickers in each row. How many stickers are on the sheet all together?

________ stickers

A game board has a grid of squares that are in 7 rows of 7 squares each. How many squares are on the game board in all?

________ squares

The desks in a classroom are arranged in 5 rows of 6 desks each. How many desks are in the classroom?

________ desks

42 Multiplication

Complete the multiplication equation shown by each number line.

EX 8 × 2 = 16

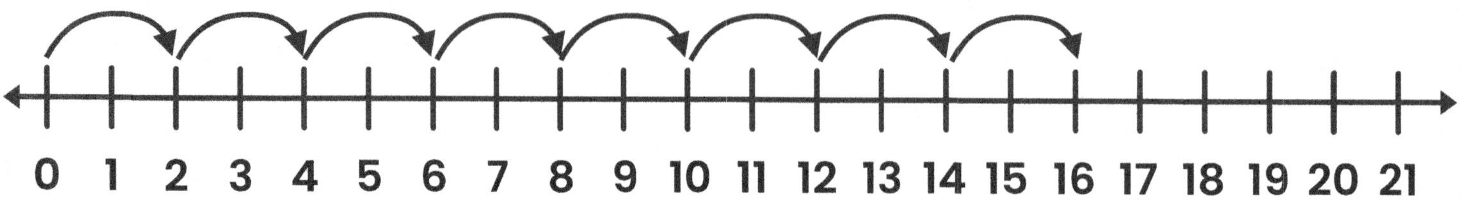

1 ______ × ______ = ______

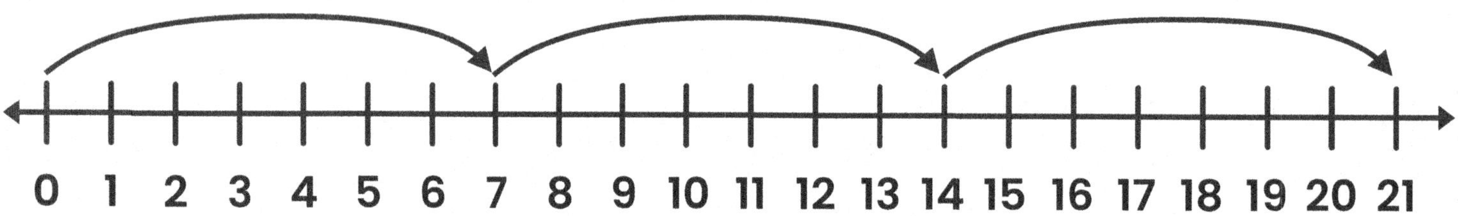

2 ______ × ______ = ______

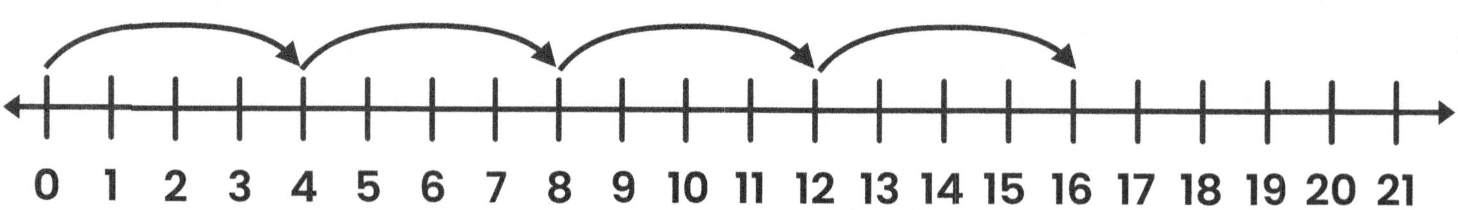

3 ______ × ______ = ______

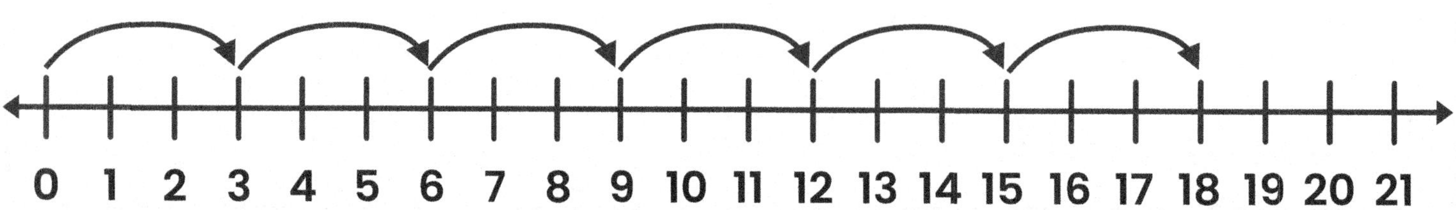

Multiplication

Show each multiplication equation on the number line.

(EX) $6 \times 2 = 12$

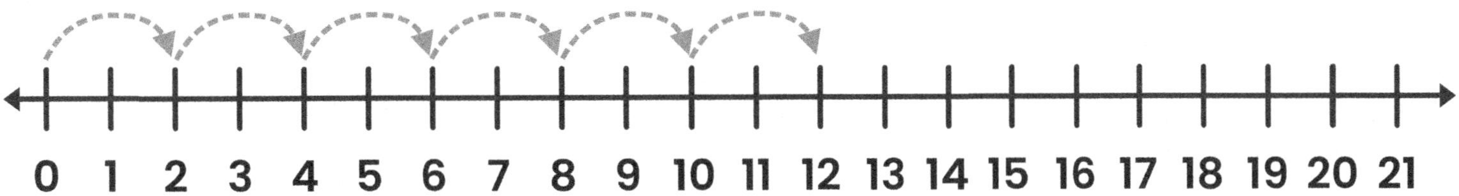

(1) $5 \times 3 = 15$

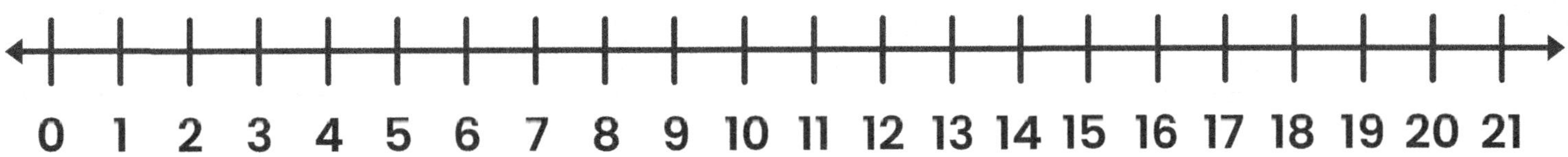

(2) $2 \times 9 = 18$

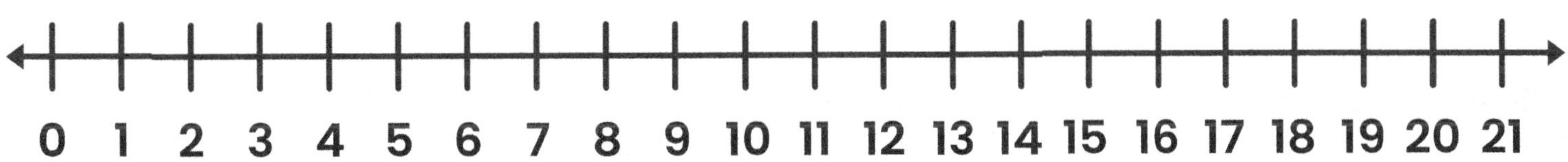

(3) $4 \times 5 = 20$

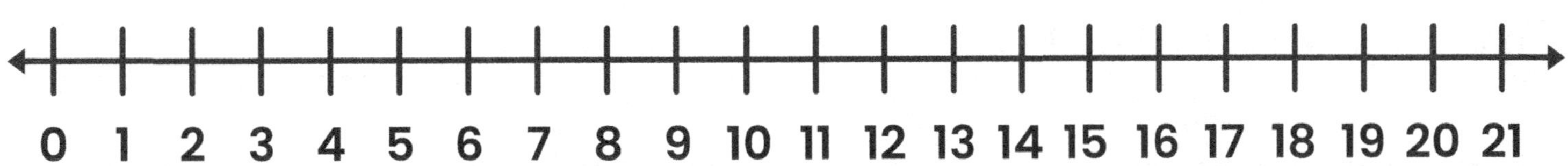

Multiplication

Multiplying by 0: Find each product.

(1) 4 × 0 = ______ (2) 5 × 0 = ______ (3) 0 × 8 = ______

(4) 0 × 7 = ______ (5) 12 × 0 = ______ (6) 11 × 0 = ______

(7) 0 × 9 = ______ (8) 10 × 0 = ______ (9) 6 × 0 = ______

(10) Write a rule for multiplying any number by 0. ______________

____

____

Multiplying by 1: Find each product.

(11) 9 × 1 = ______ (12) 3 × 1 = ______ (13) 1 × 7 = ______

(14) 1 × 8 = ______ (15) 4 × 1 = ______ (16) 6 × 1 = ______

(17) 1 × 5 = ______ (18) 10 × 1 = ______ (19) 11 × 1 = ______

(20) Write a rule for multiplying any number by 1. ______________

____

Multiplication

Multiplying by 2: Use doubles facts to multiply by 2.

(1) 2 × 2 = ______ (2) 6 × 2 = ______ (3) 2 × 3 = ______

(4) 5 × 2 = ______ (5) 2 × 7 = ______ (6) 1 × 2 = ______

(7) 2 × 4 = ______ (8) 8 × 2 = ______ (9) 2 × 9 = ______

Multiplying by 4: Use the of strategy doubling a twos fact to multiply by 4.

(EX) 4 × 6 = ?

2 × 6 = 12

12 + 12 = 24

4 × 6 = 24

(10) 4 × 7 = ?

2 × 7 = ______

______ + ______ = ______

4 × 7 = ______

(11) 4 × 4 = ?

2 × 4 = ______

______ + ______ = ______

4 × 4 = ______

(12) 4 × 8 = ?

2 × 8 = ______

______ + ______ = ______

4 × 8 = ______

46 Multiplication

Multiplying by 3: Find each product.

(1) $3 \times 2 =$ ______ (2) $6 \times 3 =$ ______ (3) $3 \times 3 =$ ______

(4) $5 \times 3 =$ ______ (5) $3 \times 7 =$ ______ (6) $1 \times 3 =$ ______

(7) $3 \times 4 =$ ______ (8) $8 \times 3 =$ ______ (9) $3 \times 9 =$ ______

Multiplying by 6: Use the of strategy doubling a threes fact to multiply by 6.

(EX) $6 \times 8 = ?$

$3 \times 8 =$ 24

24 + 24 = 48

$6 \times 8 =$ 48

(10) $6 \times 7 = ?$

$3 \times 7 =$ ______

______ + ______ = ______

$6 \times 7 =$ ______

(11) $6 \times 8 = ?$

$3 \times 8 =$ ______

______ + ______ = ______

$6 \times 8 =$ ______

(12) $6 \times 9 = ?$

$3 \times 9 =$ ______

______ + ______ = ______

$6 \times 9 =$ ______

Multiplying by 5: Count by fives to find each product.

(1) $5 \times 2 =$ ______ (2) $6 \times 5 =$ ______ (3) $5 \times 3 =$ ______

(4) $5 \times 5 =$ ______ (5) $5 \times 7 =$ ______ (6) $1 \times 5 =$ ______

(7) $4 \times 5 =$ ______ (8) $8 \times 5 =$ ______ (9) $5 \times 9 =$ ______

Multiplying by 7: Break apart the array to show how the distributive property can be used to multiply by 7.

(10) $6 \times 7 = ?$

$6 \times (2 + 5) = ?$

$(6 \times 2) + (6 \times 5) = ?$

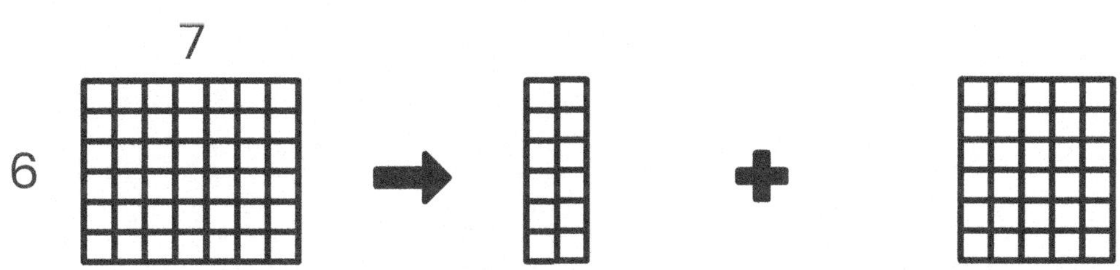

$6 \times 2 =$ ______ $6 \times 5 =$ ______

______ + ______ = ______

$6 \times 7 =$ ______

(11) $8 \times 7 = ?$

$8 \times 7 = 8 \times (2 + 5)$

$8 \times 7 = (8 \times 2) + (8 \times 5)$

$8 \times 7 =$ ______ + ______

$8 \times 7 =$ ______

(12) $9 \times 7 = ?$

$9 \times 7 = 9 \times (2 + 5)$

$9 \times 7 = (9 \times 2) + (9 \times 5)$

$9 \times 7 =$ ______ + ______

$9 \times 7 =$ ______

Multiplying by 8: Fill in the blanks to show how the distributive property can be used to multiply by 8.

(EX) $6 \times 8 = ?$

$6 \times 8 =$ 6 $\times (4 + 4)$

$6 \times 8 = (6 \times 4) + (6 \times$ 4 $)$

$6 \times 8 =$ 24 $+$ 24

$6 \times 8 =$ 48

(1) $5 \times 8 = ?$

$5 \times 8 = 5 \times (3 +$ ______ $)$

$5 \times 8 = (5 \times 3) + (5 \times$ ______ $)$

$5 \times 8 =$ ______ $+$ ______

$5 \times 8 =$ ______

(2) $7 \times 8 = ?$

$7 \times 8 =$ ______ $\times (3 + 5)$

$7 \times 8 = (7 \times 3) + (7 \times$ ______ $)$

$7 \times 8 =$ ______ $+$ ______

$7 \times 8 =$ ______

(3) $9 \times 8 = ?$

$9 \times 8 = 9 \times (4 +$ ______ $)$

$9 \times 8 = (9 \times 4) + (9 \times$ ______ $)$

$9 \times 8 =$ ______ $+$ ______

$9 \times 8 =$ ______

Use the associative property to multiply by 8.

(EX) $8 \times 4 = ?$

$8 \times 4 =$ 4 $\times 8$

4 $\times 8 =$ 32

So, $8 \times 4 =$ 32

(4) $8 \times 3 = ?$

$8 \times 3 =$ ______ $\times 8$

______ $\times 8 =$ ______

So, $8 \times 3 =$ ______

(5) $8 \times 2 = ?$

$8 \times 2 =$ ______ $\times 8$

______ $\times 8 =$ ______

So, $8 \times 2 =$ ______

(6) $8 \times 5 = ?$

$8 \times 5 =$ ______ $\times 8$

______ ______

So, $8 \times 5 =$ ______

Multiplying by 9: Fill in the multiples of 9 in the multiplication table.

×	1	2	3	4	5	6	7	8	9
1									
2									
3									
4									
5									
6									
7									
8									
9									

Describe a pattern you see in the multiples of 9. ____________________

____

____

____

Multiplication

Multiplying by 10: Find each product.

(1) $10 \times 1 =$ ______

(2) $4 \times 10 =$ ______

(3) $10 \times 7 =$ ______

(4) $5 \times 10 =$ ______

(5) $10 \times 3 =$ ______

(6) $8 \times 10 =$ ______

(7) $10 \times 6 =$ ______

(8) $9 \times 10 =$ ______

(9) $2 \times 10 =$ ______

(10) $10 \times 10 =$ ______

(11) $10 \times 12 =$ ______

(12) $24 \times 10 =$ ______

(13) $32 \times 10 =$ ______

(14) $10 \times 61 =$ ______

(15) $10 \times 80 =$ ______

(16) $55 \times 10 =$ ______

(17) $17 \times 10 =$ ______

(18) $10 \times 74 =$ ______

(19) $10 \times 28 =$ ______

(20) Write a rule for multiplying any number by 10. ____________________

____

Find each product.

2	8	3	4	5	8	6
× 2	× 2	× 7	× 6	× 5	× 5	× 6
6	3	2	8	7	5	5
× 7	× 2	× 9	× 3	× 4	× 6	× 9
7	8	2	3	3	4	7
× 9	× 6	× 4	× 2	× 9	× 8	× 5
1	8	6	5	4	4	9
× 9	× 8	× 9	× 2	× 3	× 4	× 4
2	4	8	7	2	3	4
× 0	× 1	× 9	× 7	× 6	× 5	× 5
0	1	4	9	8	7	6
× 6	× 7	× 7	× 9	× 7	× 2	× 3

Multiplication

Complete the multiplication table.

×	1	2	3	4	5	6	7	8	9	10
1										
2										
3										
4										
5										
6										
7										
8										
9										
10										

Which of these will always give you an even product?
(select all that apply)

○ even × even ○ even × odd ○ odd × odd

Which of these will always give you an odd product?
(select all that apply)

○ even × even ○ even × odd ○ odd × odd

Multiplication

Fill the the blanks with the correct number to complete each sentence.

1. 16 is 2 times as much as ______ .
2. ______ is 4 times as much as 3.
3. 18 is ______ times as much as 9.
4. 24 is 4 times as much as ______ .
5. ______ is 3 times as much as 5.
6. 28 is ______ times as much as 7.
7. 56 is 7 times as much as ______ .
8. ______ is 6 times as much as 8.

Write an equation to represent each sentence.

9. 72 is 8 times as much as 9. ______ × ______ = ______
10. 30 is 10 times as much as 3. ______ × ______ = ______
11. 14 is 7 times as much as 2. ______ × ______ = ______
12. 35 is 5 times as much as 7. ______ × ______ = ______
13. 27 is 9 times as much as 3. ______ × ______ = ______
14. 36 is 4 times as much as 9. ______ × ______ = ______

Multiplication

Circle whether each equation is true or false.

1. TRUE or FALSE ?

   $6 \times 3 = 24$

2. TRUE or FALSE ?

   $3 \times 7 = 21$

3. TRUE or FALSE ?

   $8 \times 5 = 45$

4. TRUE or FALSE ?

   $7 \times 4 = 27$

5. TRUE or FALSE ?

   $9 \times 8 = 72$

6. TRUE or FALSE ?

   $8 \times 8 = 64$

Fill in the missing number that would make each equation true.

7. 9 × 3 = ____

8. 5 × ____ = 40

9. 3 × ____ = 30

10. ____ × 6 = 48

11. 4 × 9 = ____

12. ____ × 4 = 0

13. 9 × ____ = 81

14. ____ × 7 = 56

15. 2 × 8 = ____

16. ____ × 28 = 28

Use multiplication to solve each word problem.

1. Mr. Franklin gives his students a weekly spelling test. There are 10 words on each week's spelling test. How many words will the students have been tested on at the end of 8 weeks?

   ______ × ______ = ______ words

2. Lola brings 4 packages of cupcakes to school to share with her class. There are 6 cupcakes in each package. How many cupcakes does Lola bring?

   ______ × ______ = ______ cupcakes

3. A baseball game lasts 9 innings unless the game is tied. If none of the games were tied, how many total innings would be played in 8 baseball games?

   ______ × ______ = ______ innings

4. Ellen has 8 place settings of silverware. Each place setting contains a fork, a spoon, and a knife. How many pieces of silverware does Ellen have in all?

   ______ × ______ = ______ pieces of silverware

5. Each of the 3 bedrooms in Rosa's house has one window. If Rosa wants to hang 2 curtains on every bedroom window, how many will she need?

   ______ × ______ = ______ curtains

Multiplication

Fill in the blanks to make each equation true.

1. $7 \times 8 =$ ______ $\times 7$

2. $6 \times 9 = (6 \times 5) + (6 \times$ ______ $)$

3. $3 \times (5 \times 6) = (3 \times$ ______ $) \times 6$

4. $8 \times 6 = 8 \times 2 \times$ ______

5. $2 \times 7 \times 4 = 7 \times$ ______ $\times 2$

6. $5 \times 7 = ($ ______ $\times 2) + (5 \times 5)$

7. $(4 \times 2) \times 9 = 4 \times ($ ______ $\times 9)$

8. $7 \times 9 = 7 \times 3 \times$ ______

9. $3 \times 4 \times 9 = 4 \times 9 \times$ ______

10. $3 \times 8 = (3 \times 2) + (3 \times$ ______ $)$

Fill in the correct bubble to answer each question.

Which of these is equal to 7×6?

- ○ 6×7
- ○ $7 \times 3 \times 3$
- ○ $4 + (3 \times 6)$

Which of these is equal to 8×3?

- ○ $4 \times 4 \times 3$
- ○ $3 + 8$
- ○ $(4 \times 3) + (4 \times 3)$

Which of these is equal to $2 \times (3 \times 4)$?

- ○ 2×7
- ○ $(2 \times 3) \times 4$
- ○ $4 \times (2 + 3)$

Which of these is equal to $5 \times 3 \times 2$?

- ○ 5×5
- ○ 5×6
- ○ 8×2

Which of these is equal to $4 \times 5 \times 6$?

- ○ $5 \times 6 \times 4$
- ○ 9×6
- ○ $4 \times (5 + 6)$

Multiplication

Write all of the factor pairs that multiply to the product shown in each box.

12
1 × 12
2 × 6
3 × 4
4 × 3
6 × 2
12 × 1

8

24

6

16

7

Circle whether each equation is true or false.

1. **TRUE or FALSE ?** $8 \times 3 = 4 \times 6$

2. **TRUE or FALSE ?** $2 \times 7 = 3 \times 5$

3. **TRUE or FALSE ?** $8 \times 3 = 9 \times 2$

4. **TRUE or FALSE ?** $7 \times 3 = 4 \times 5$

5. **TRUE or FALSE ?** $9 \times 4 = 6 \times 6$

6. **TRUE or FALSE ?** $1 \times 27 = 9 \times 3$

Fill in any numbers that would make each equation true.

7. $4 \times 3 =$ ____ × ____

8. $7 \times 2 =$ ____ × ____

9. $3 \times 5 =$ ____ × ____

10. $3 \times 6 =$ ____ × ____

11. $4 \times 5 =$ ____ × ____

12. $6 \times 4 =$ ____ × ____

Multiplication

Multiples of 10: Find each product.

(EX) $3 \times 80 = 3 \times$ 8 tens
$3 \times 80 =$ 24 tens
$3 \times 80 =$ 240

(1) $4 \times 50 = 4 \times$ ______ tens
$4 \times 50 =$ ______ tens
$4 \times 50 =$ ______

(2) $7 \times 90 = 7 \times$ ______ tens
$7 \times 90 =$ ______ tens
$7 \times 90 =$ ______

(3) $5 \times 60 = 5 \times$ ______ tens
$5 \times 60 =$ ______ tens
$5 \times 60 =$ ______

(4) $8 \times 40 = 8 \times$ ______ tens
$8 \times 40 =$ ______ tens
$8 \times 40 =$ ______

(5) $9 \times 20 = 9 \times$ ______ tens
$9 \times 20 =$ ______ tens
$9 \times 20 =$ ______

(6) $2 \times 70 =$ ______

(7) $6 \times 40 =$ ______

(8) $3 \times 90 =$ ______

(9) $4 \times 30 =$ ______

(10) Write a rule for multiplying a 1-digit number by a multiple of 10.

____

____

Use multiplication to solve each word problem.

1. There are 8 students on a STEM Club team. They each plan to buy matching sweatshirts that cost $20. What will be the total cost of the team's sweatshirts?

   ______ × ______ = ______ dollars

2. Maple Elementary School includes kindergarten through fifth grade. If there are 90 students in each grade, how many students attend the school?

   ______ × ______ = ______ students

3. Pencils come in boxes of 40 pencils each. How many pencils are in 7 boxes?

   ______ × ______ = ______ pencils

4. Amanda practices playing the flute for 30 minutes a day, 5 days a week. How many minutes does she spend practicing each week?

   ______ × ______ = ______ minutes

5. Joe's goal is to read 80 pages a week. If he meets his goal, how many pages will he have read at the end of 6 weeks?

   ______ × ______ = ______ pages

Division

Fill in the blanks to describe each picture.

EX

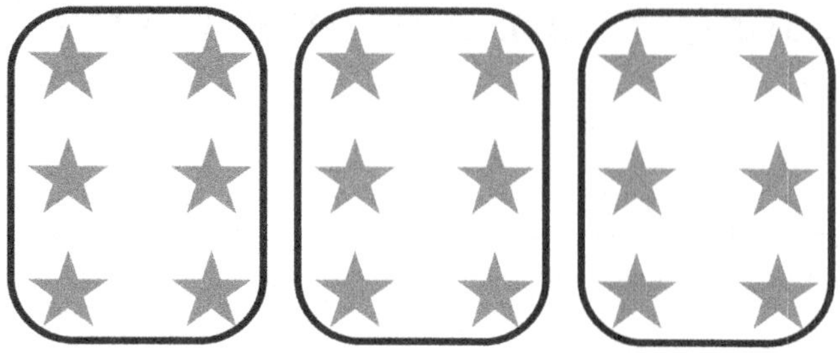

There are 18 stars divided into 3 equal groups.

There are __6__ stars in each group.

So, 18 ÷ 3 = __6__.

1

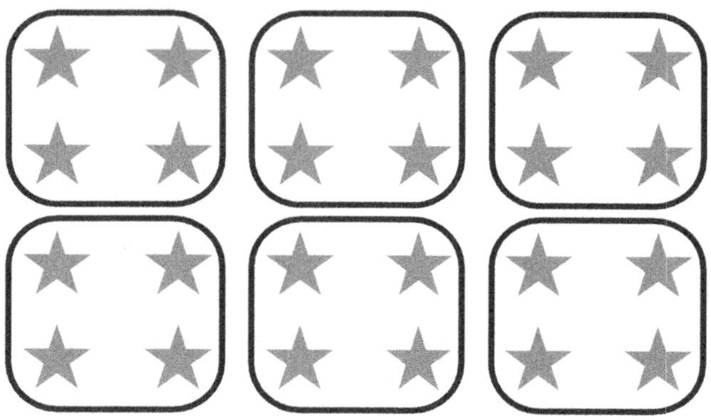

There are 24 stars divided into 6 equal groups.

There are ______ stars in each group.

So, 24 ÷ 6 = ______ .

2

There are 15 stars divided into 3 equal groups.

There are ______ stars in each group.

So, 15 ÷ 3 = ______.

3

There are 12 stars divided into 4 equal groups.

There are ______ stars in each group.

So, 12 ÷ 4 = ______.

Fill in the blanks to describe each picture.

EX

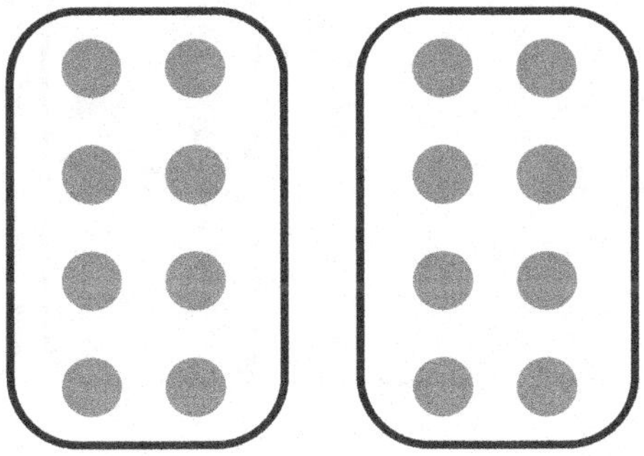

There are __16__ dots.

They are divided into __2__ equal groups.

There are __8__ dots in each group.

So, __16__ ÷ __2__ = __8__.

1

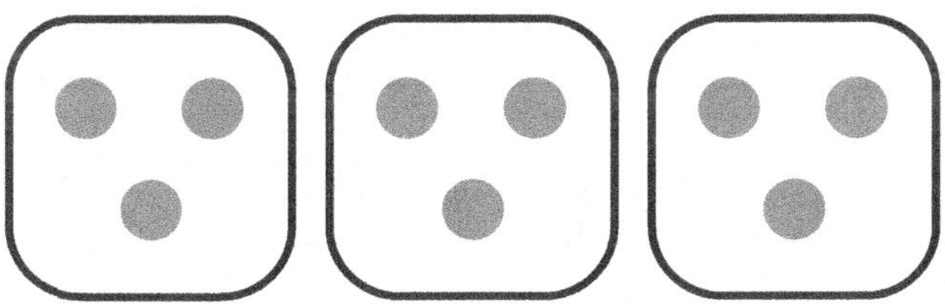

There are ______ dots.

They are divided into ______ equal groups.

There are ______ dots in each group.

So, ______ ÷ ______ = ______.

2

There are ______ dots.

They are divided into ______ equal groups.

There are ______ dots in each group.

So, ______ ÷ ______ = ______.

3

There are ______ dots.

They are divided into ______ equal groups.

There are ______ dots in each group.

So, ______ ÷ ______ = ______.

Division

Write a division equation to describe each picture.

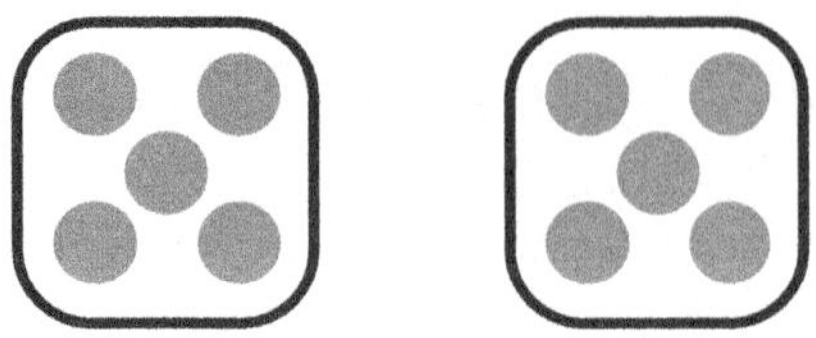

10 ÷ 2 = 5

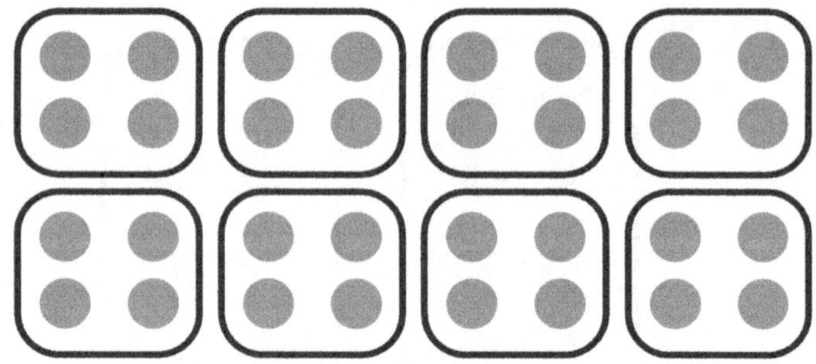

______ ÷ ______ = ______

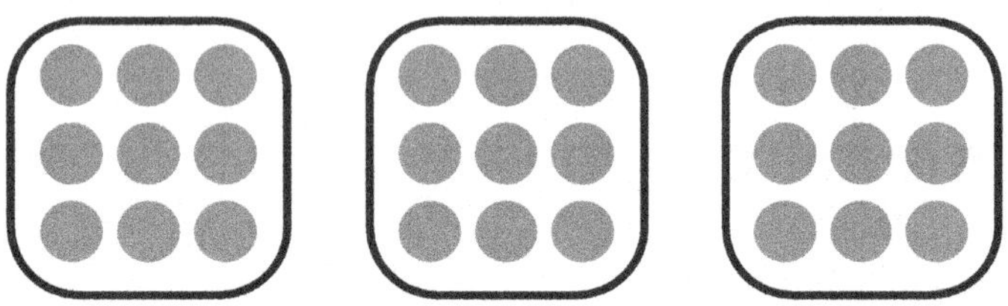

______ ÷ ______ = ______

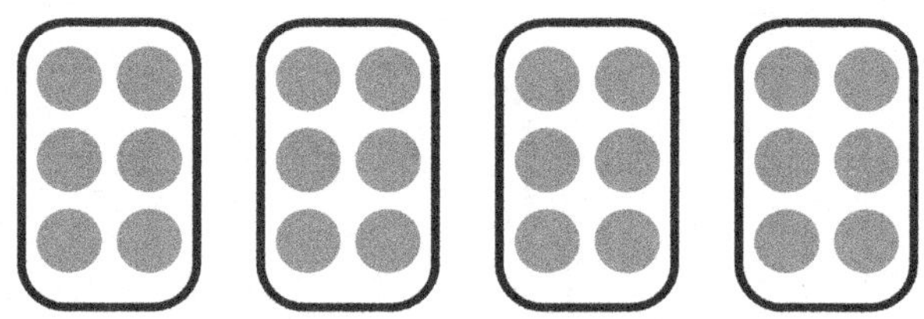

______ ÷ ______ = ______

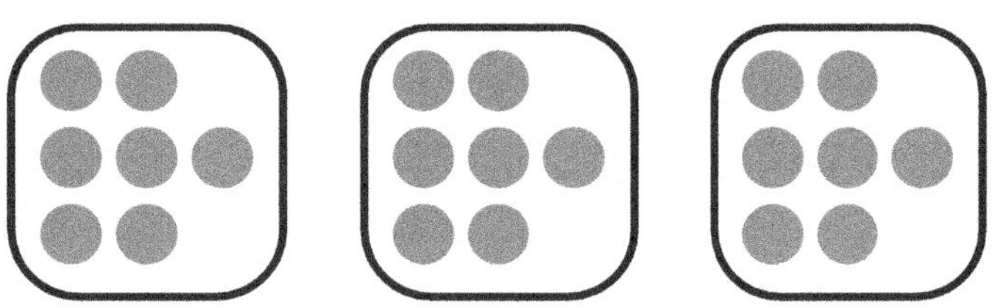

______ ÷ ______ = ______

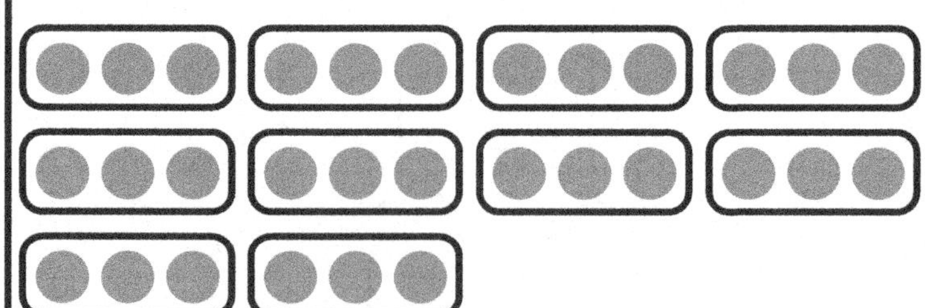

______ ÷ ______ = ______

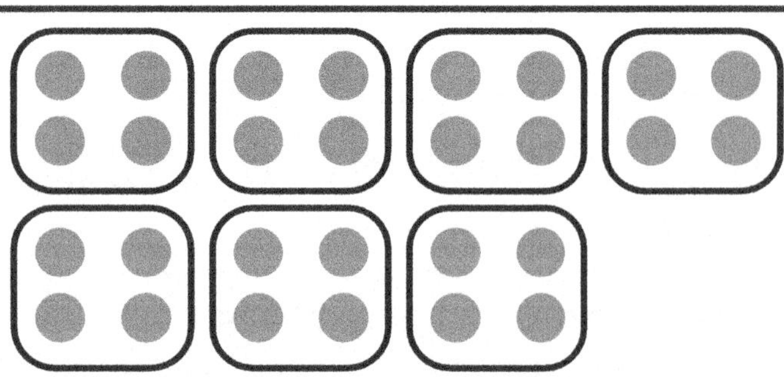

______ ÷ ______ = ______

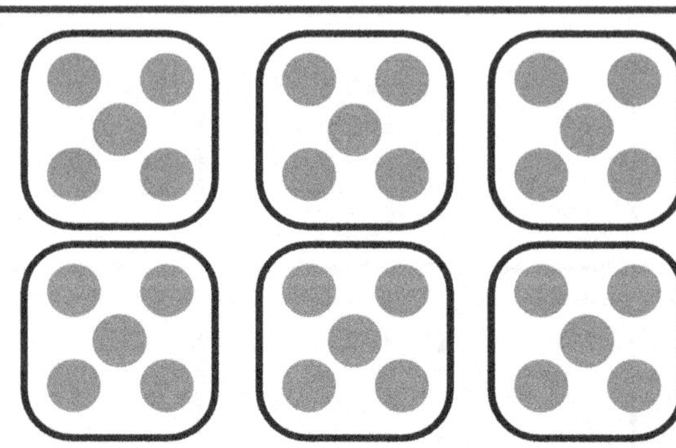

______ ÷ ______ = ______

Draw a picture to represent each division equation, then solve.

$6 \div 3 =$ _____	$14 \div 7 =$ _____
$24 \div 3 =$ _____	$36 \div 6 =$ _____
$28 \div 7 =$ _____	$12 \div 2 =$ _____
$16 \div 4 =$ _____	$25 \div 5 =$ _____

Division

Draw equal groups to help you solve each division word problem.

A class of 24 students splits up into 6 equal groups for an activity. How many students are in each group?

________ students

Mary is making bracelets. She needs 8 beads for each bracelet. If the beads come in packs of 72, how many bracelets can she make with one pack?

________ bracelets

Glenn plants 42 carrot seeds in 7 equal rows in his garden. How many carrot seeds are in each row?

________ carrot seeds

Draw equal groups to represent each word problem. Write a division equation and solve.

A pizza is cut into 12 slices. If 6 people share it equally, how many slices will each person get?

______ ÷ ______ = ______ slices

If 5 friends equally share 45 marbles, how many will each one get?

______ ÷ ______ = ______ marbles

Alex has uses 36 croutons on 4 salads. How many croutons does he put on each salad?

______ ÷ ______ = ______ croutons

Division

Complete the division equation shown by each number line.

EX 18 ÷ 2 = 9

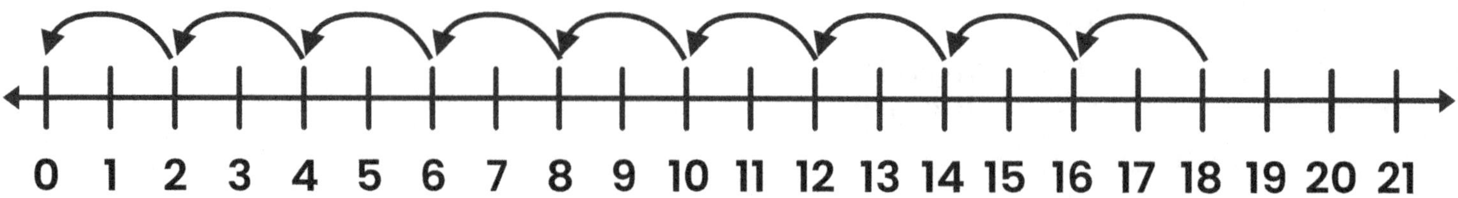

1. ______ ÷ ______ = ______

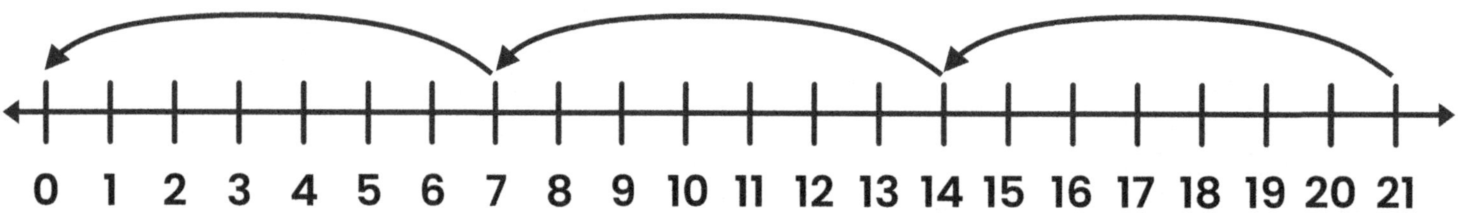

2. ______ ÷ ______ = ______

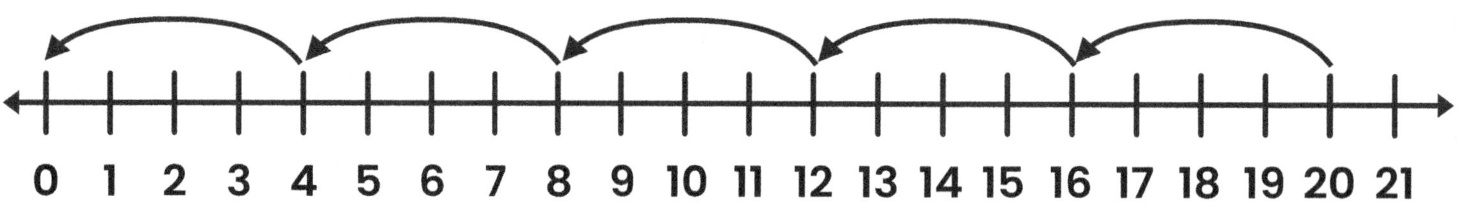

3. ______ ÷ ______ = ______

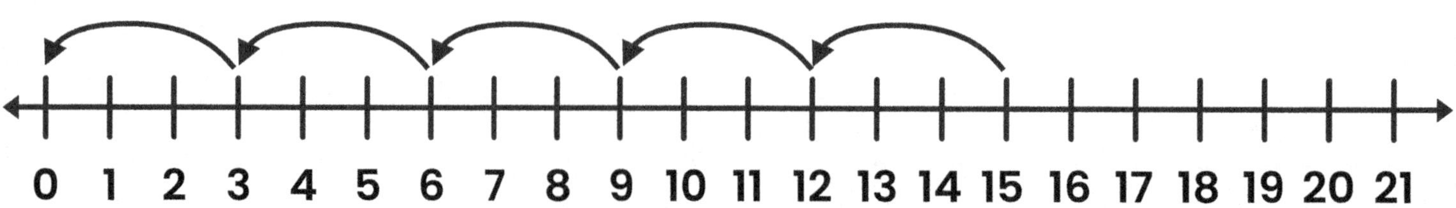

Show each division equation on the number line.

(EX) $10 \div 2 = 5$

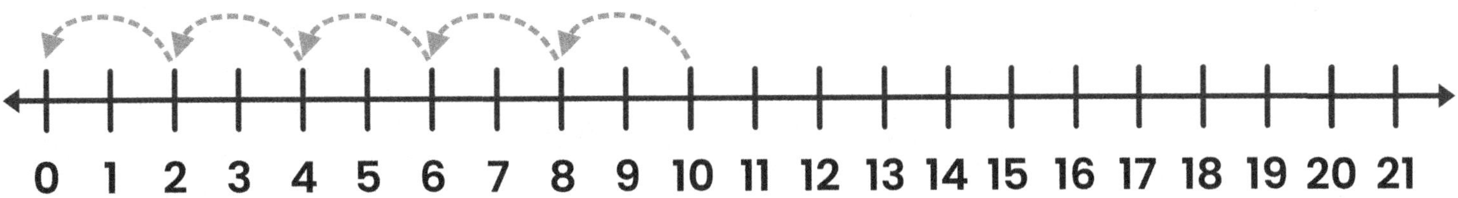

(1) $16 \div 4 = 4$

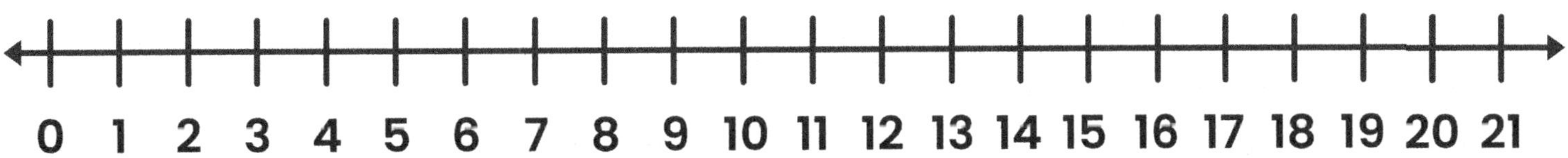

(2) $8 \div 4 = 2$

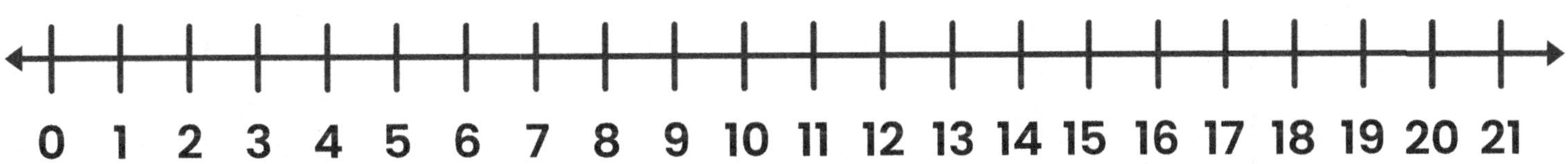

(3) $14 \div 2 = 7$

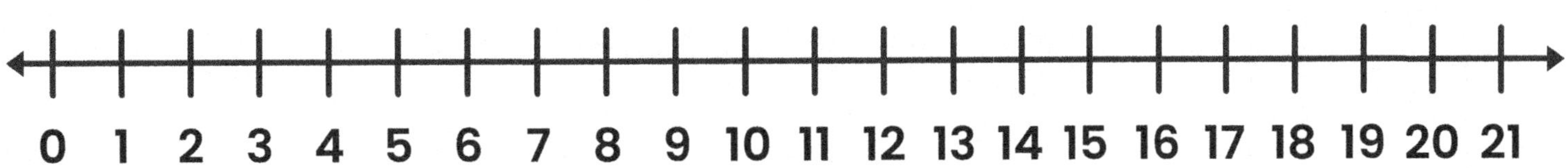

Division

Fill in the blanks to describe each array.

(EX) 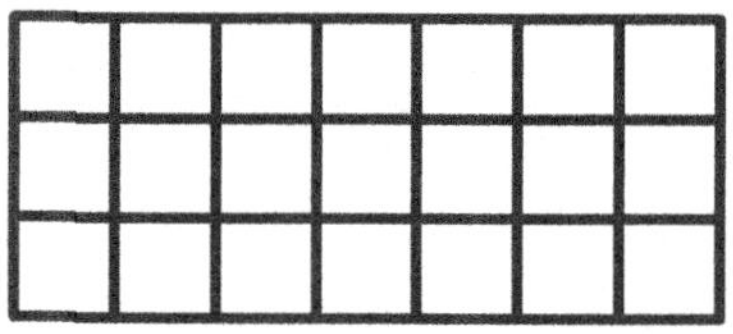

There are 21 squares in __3__ equal rows.

There are __7__ squares in each row.

So, 21 ÷ __3__ = __7__.

(1) 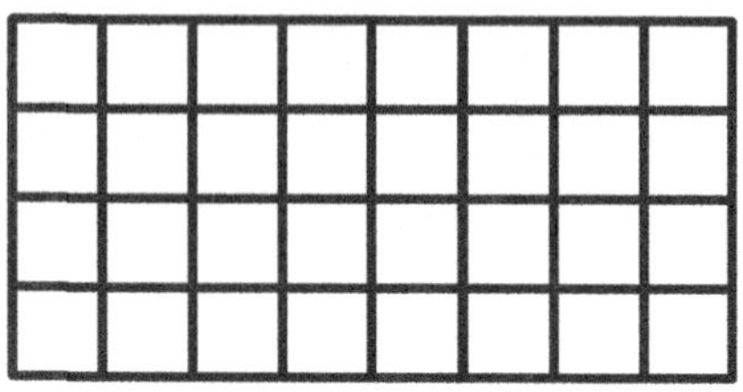

There are 32 squares in ______ equal rows.

There are ______ squares in each row.

So, 32 ÷ ______ = ______.

(2) 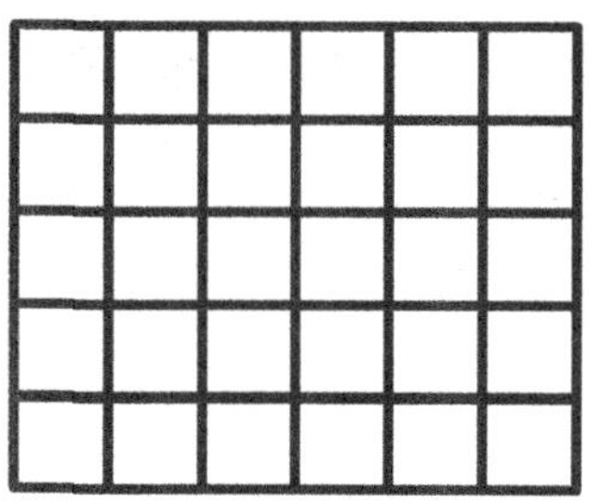

There are 30 squares in ______ equal rows.

There are ______ squares in each row.

So, 30 ÷ ______ = ______.

(3) 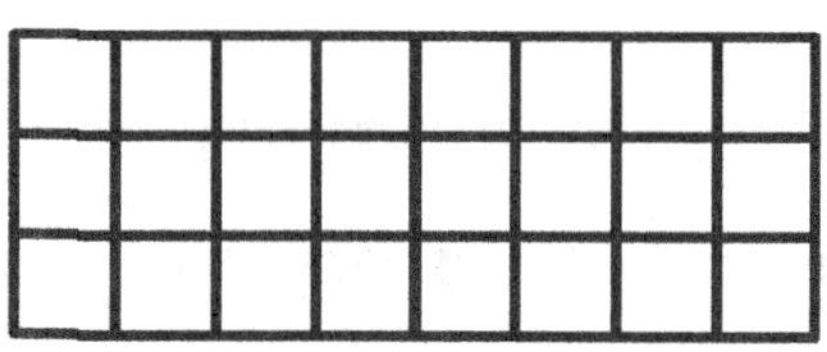

There are 24 squares in ______ equal rows.

There are ______ squares in each row.

So, 24 ÷ ______ = ______.

(4) 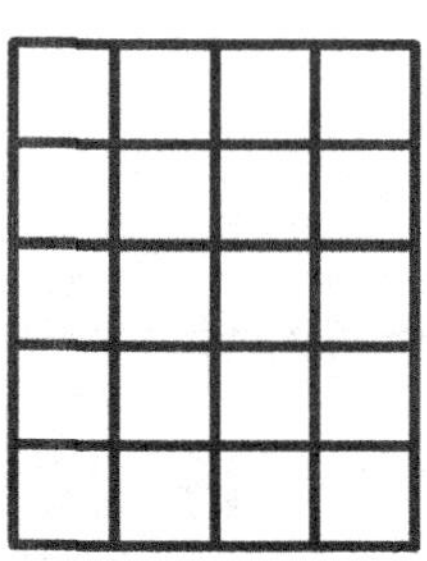

There are 20 squares in ______ equal rows.

There are ______ squares in each row.

So, 20 ÷ ______ = ______.

Complete the multiplication equation by writing the missing factor. Then, use it to divide.

(EX) 7 × 8 = 56
56 ÷ 7 = 8

(1) 5 × ______ = 35
35 ÷ 5 = ______

(2) ______ × 9 = 27
27 ÷ 9 = ______

(3) ______ × 7 = 14
14 ÷ 7 = ______

(4) 8 × ______ = 64
64 ÷ 8 = ______

(5) 6 × ______ = 30
30 ÷ 6 = ______

(6) ______ × 2 = 18
18 ÷ 2 = ______

(7) ______ × 3 = 24
24 ÷ 3 = ______

(8) 4 × ______ = 36
36 ÷ 4 = ______

(9) 2 × ______ = 16
16 ÷ 2 = ______

(10) ______ × 3 = 6
6 ÷ 3 = ______

(11) ______ × 5 = 25
25 ÷ 5 = ______

(12) 7 × ______ = 21
21 ÷ 7 = ______

(13) 9 × ______ = 72
72 ÷ 9 = ______

Division

Dividing with 0: Find each quotient.

1. $0 \div 4 =$ ______
2. $0 \div 5 =$ ______
3. $0 \div 8 =$ ______
4. $0 \div 7 =$ ______
5. $0 \div 9 =$ ______
6. $0 \div 3 =$ ______
7. Write a rule for dividing 0 by any other number. ______

______

8. Does dividing by 0 result in the same answer? Why or why not?

______

______

Dividing by 1: Find each quotient.

9. $9 \div 1 =$ ______
10. $3 \div 1 =$ ______
11. $7 \div 1 =$ ______
12. $8 \div 1 =$ ______
13. $4 \div 1 =$ ______
14. $6 \div 1 =$ ______
15. Write a rule for dividing any number by 1. ______

______

16. Draw a picture to show why this is true.

Dividing a Number by Itself: Find each quotient.

(1) $2 \div 2 =$ ______ (2) $5 \div 5 =$ ______ (3) $7 \div 7 =$ ______

(4) $1 \div 1 =$ ______ (5) $6 \div 6 =$ ______ (6) $9 \div 9 =$ ______

(7) Write a rule for dividing any number greater than 0 by itself.

____

____

(8) Draw a picture to show why this is true.

Dividing by 2: Find each quotient.

(9) $10 \div 2 =$ ______ (10) $6 \div 2 =$ ______ (11) $12 \div 2 =$ ______

(12) $16 \div 2 =$ ______ (13) $14 \div 2 =$ ______ (14) $2 \div 2 =$ ______

(15) $8 \div 2 =$ ______ (16) $18 \div 2 =$ ______ (17) $4 \div 2 =$ ______

Division

Dividing by 3: Find each quotient.

(1) $3 \div 3 =$ ______ (2) $9 \div 3 =$ ______ (3) $15 \div 3 =$ ______

(4) $24 \div 3 =$ ______ (5) $27 \div 3 =$ ______ (6) $18 \div 3 =$ ______

(7) $21 \div 3 =$ ______ (8) $12 \div 3 =$ ______ (9) $6 \div 3 =$ ______

Dividing by 4: Find each quotient.

(10) $36 \div 4 =$ ______ (11) $32 \div 4 =$ ______ (12) $12 \div 4 =$ ______

(13) $20 \div 4 =$ ______ (14) $24 \div 4 =$ ______ (15) $8 \div 4 =$ ______

(16) $28 \div 4 =$ ______ (17) $16 \div 4 =$ ______ (18) $4 \div 4 =$ ______

Dividing by 5: Find each quotient.

(19) $20 \div 5 =$ ______ (20) $30 \div 5 =$ ______ (21) $5 \div 5 =$ ______

(22) $25 \div 5 =$ ______ (23) $35 \div 5 =$ ______ (24) $15 \div 5 =$ ______

(25) $40 \div 5 =$ ______ (26) $45 \div 5 =$ ______ (27) $10 \div 5 =$ ______

Dividing by 6: Find each quotient.

(1) 54 ÷ 6 = ______ (2) 48 ÷ 6= ______ (3) 12 ÷ 6 = ______

(4) 24 ÷ 6 = ______ (5) 36 ÷ 6 = ______ (6) 18 ÷ 6 = ______

(7) 30÷ 6 = ______ (8) 42 ÷ 6 = ______ (9) 6 ÷ 6 = ______

Dividing by 7: Find each quotient.

(10) 35 ÷ 7 = ______ (11) 42 ÷ 7 = ______ (12) 7 ÷ 7 = ______

(13) 28 ÷ 7 = ______ (14) 49 ÷ 7 = ______ (15) 21 ÷ 7 = ______

(16) 56 ÷ 7 = ______ (17) 63 ÷ 7 = ______ (18) 14 ÷ 7 = ______

Dividing by 8: Find each quotient.

(19) 24 ÷ 8 = ______ (20) 56 ÷ 8 = ______ (21) 16 ÷ 8 = ______

(22) 48 ÷ 8 = ______ (23) 32 ÷ 8 = ______ (24) 8 ÷ 8 = ______

(25) 64 ÷ 8 = ______ (26) 40 ÷ 8 = ______ (27) 72 ÷ 8 =______

Division

Dividing by 9: Find each quotient.

1. $36 \div 9 =$ ______
2. $72 \div 9 =$ ______
3. $81 \div 9 =$ ______
4. $27 \div 9 =$ ______
5. $54 \div 9 =$ ______
6. $9 \div 9 =$ ______
7. $63 \div 9 =$ ______
8. $45 \div 9 =$ ______
9. $18 \div 9 =$ ______

Mixed Review: Find each quotient.

10. $27 \div 3 =$ ______
11. $32 \div 4 =$ ______
12. $8 \div 1 =$ ______
13. $24 \div 6 =$ ______
14. $35 \div 5 =$ ______
15. $18 \div 2 =$ ______
16. $63 \div 7 =$ ______
17. $54 \div 9 =$ ______
18. $16 \div 8 =$ ______
19. $3 \div 1 =$ ______
20. $24 \div 3 =$ ______
21. $14 \div 2 =$ ______
22. $28 \div 4 =$ ______
23. $36 \div 6 =$ ______
24. $21 \div 7 =$ ______
25. $40 \div 5 =$ ______
26. $56 \div 8 =$ ______
27. $81 \div 9 =$ ______

Division

Fill in the correct bubble to answer each question.

Which of these is equal to $64 \div 8$?

- ○ 6
- ○ 8
- ○ 9

Which of these is equal to 7?

- ○ $45 \div 5$
- ○ $54 \div 8$
- ○ $42 \div 6$

Which of these has the same quotient as $12 \div 3$?

- ○ $24 \div 8$
- ○ $27 \div 3$
- ○ $28 \div 7$

Which of these has the same quotient as $36 \div 4$?

- ○ $63 \div 7$
- ○ $56 \div 7$
- ○ $49 \div 7$

Which of these has the same quotient as $16 \div 2$?

- ○ $42 \div 7$
- ○ $40 \div 5$
- ○ $24 \div 4$

Division

Circle whether each equation is true or false.

① TRUE or FALSE ?

$56 \div 7 = 8$

② TRUE or FALSE ?

$63 \div 9 = 6$

③ TRUE or FALSE ?

$48 \div 8 = 6$

④ TRUE or FALSE ?

$28 \div 3 = 9$

⑤ TRUE or FALSE ?

$45 \div 5 = 8$

⑥ TRUE or FALSE ?

$49 \div 7 = 7$

Fill in the missing number that would make each equation true.

⑦ $9 \div 3 =$ ____

⑧ $40 \div$ ____ $= 5$

⑨ $32 \div$ ____ $= 4$

⑩ ____ $\div 6 = 7$

⑪ $54 \div 9 =$ ____

⑫ ____ $\div 8 = 8$

⑬ $12 \div$ ____ $= 2$

⑭ ____ $\div 3 = 6$

⑮ $36 \div 6 =$ ____

⑯ ____ $\div 2 = 7$

Use division to solve each problem.

1. Alex buys 40 pounds of dog food. There are 5 pounds in each bag. How many bags of dog food does Alex buy?

   ______ ÷ ______ = ______ bags

2. 63 pretzels are divided evenly among 7 people. How many pretzels does each person get?

   ______ ÷ ______ = ______ pretzels

3. Sam has 81 baseball cards that he is putting into protective sheets. Each sheet has enough pockets to hold 9 cards. How many sheets will Sam use to hold all of his cards?

   ______ ÷ ______ = ______ sheets

4. Megan is baking rolls that come in a package of 8. If she wants to make 64 rolls all together, how many packages will she need to use?

   ______ ÷ ______ = ______ packages

5. Shawna makes a rectangular tray of brownies and slices them into 3 rows. If there are 24 brownies all together, how many are in each row?

   ______ ÷ ______ = ______ brownies

Mixed Operations

For each set of three numbers given, write four related multiplication and division facts.

1.

3	18	6

____ × ____ = ____

____ × ____ = ____

____ ÷ ____ = ____

____ ÷ ____ = ____

2.

28	4	7

____ × ____ = ____

____ × ____ = ____

____ ÷ ____ = ____

____ ÷ ____ = ____

Multiply or divide.

3. 21 ÷ 3 = ____
4. 9 × 5 = ____
5. 81 ÷ 9 = ____
6. 56 ÷ 7 = ____
7. 6 × 4 = ____
8. 36 ÷ 6 = ____
9. 8 × 8 = ____
10. 49 ÷ 7 = ____
11. 7 × 10 = ____
12. 40 ÷ 10 = ____
13. 3 × 8 = ____
14. 27 ÷ 3 = ____

Multiply or divide to solve each word problem.

1. Juice boxes come in packs of 6. A team needs 48 juice boxes. How many packs do they need?

   ______ packs

2. Paula is making omelets using 3 eggs each. How many omelets can she make if she uses a whole carton of 12 eggs?

   ______ omelets

3. Damian has 36 books. If he splits them evenly between 4 shelves, how many books will be on each shelf?

   ______ books

4. Reagan's cookie recipe calls for 2 cups of flour per batch. How many cups of flour would Reagan need to make 5 batches?

   ______ cups

5. Paper plates come in packs of 8. If there will be 56 guests at a party, how many packs of plates are needed to give each guest one plate?

   ______ packs

Mixed Operations

In each equation, a letter stands for an unknown number. Find the value of the unknown number, then write a related fact to show how you know.

(EX) $7 \times \mathbf{n} = 21$

n = 3

Related fact:

21 ÷ 7 = 3

(1) $12 + \mathbf{p} = 20$

p = ______

Related fact:

____

(2) $35 \div \mathbf{m} = 5$

m = ______

Related fact:

____

(3) $25 - \mathbf{a} = 19$

a = ______

Related fact:

____

(4) $\mathbf{b} + 8 = 17$

b = ______

Related fact:

____

(5) $\mathbf{q} \times 9 = 54$

q = ______

Related fact:

____

(6) $\mathbf{d} - 3 = 14$

d = ______

Related fact:

____

(7) $\mathbf{c} \div 4 = 7$

c = ______

Related fact:

____

Solve each word problem using the correct operation.

1. Tara has $78. She spends $14 on a bracelet. How much money does she have left?

2. A can a soda contains 8 ounces. How many ounces of soda are in a six-pack of cans?

3. On Monday, Seth runs for 3 kilometers. On Tuesday, he runs for 3 more kilometers than he did on Monday. How many kilometers did he run in total on both days?

4. Fiona has a goal of drinking 72 ounces of water while she's at work each day. If she works for 8 hours a day, how many ounces should she try to drink per hour to meet her goal?

5. Reggie earns $30 an hour at his job. How much will he earn after working 9 hours?

Mixed Operations

Fill in both equations to solve each two-step word problem.

(EX) Kim is making quiches for a brunch. She has a full carton of 12 eggs and another carton that contains 8 eggs. Each quiche takes 4 eggs, and she uses all the eggs. How many quiches does Kim make?

12 + 8 = 20 eggs

20 ÷ 4 = 5 quiches

(1) Damian took a box of crayons and gave 4 to each of the 8 kids in his group. When he finished, there were 7 crayons left in the box. How many crayons were originally in the box?

______ × ______ = ______ crayons given out

______ + ______ = ______ crayons originally in the box

(2) Marcus was carrying a stack of 12 plates when he tripped. Some of the plates fell and 2 of them broke. He set down the remaining plates in 2 equal piles. How many plates were in each pile?

______ − ______ = ______ remaining plates

______ ÷ ______ = ______ plates in each pile

3. A book has 1 chapter that is 12 pages long and the other 8 chapters are each 10 pages long. How many pages are in the book all together?

______ × ______ = ______ pages in the equal chapters

______ + ______ = ______ pages

4. There are 26 girls and 22 boys in a P.E. class. They split up into teams of 8 students for a game. How many teams are there?

______ + ______ = ______ students in the class

______ ÷ ______ = ______ teams

5. At a Thanksgiving dinner, there are 4 pies, each cut into 8 slices. After 3 people each eat a slice, how many slices are left?

______ × ______ = ______ slices of pie in all

______ – ______ = ______ slices left

6. Jamie is 7 years old. Jamie's brother Craig is 3 years older than Jamie. Their mom is 4 times as old as Craig. How old is their mom?

______ + ______ = ______ years old (Craig's age)

______ × ______ = ______ years old (their mom's age)

Mixed Operations

Write two equations to solve each word problem.

1. Keisha has 7 five-dollar bills. She spends $12 on a toy. How much money does she have left?

2. Aiden has a sheet of stickers that contains 6 rows of 4 stickers. After he uses 5 of the stickers, how many are left?

3. A party venue charges a flat rate of $200, plus $9 per party guest. What would it cost to have a party there for 10 people?

4. There are 9 kids at a park. Then, 3 more kids come. They split into 2 teams for a game. How many kids are on each team?

5. Aria buys one 8-foot wooden board and four 6-foot boards. What is the total length of the wooden boards she buys?

6. Josie cuts a 36-inch-long ribbon into 4 equal pieces. Then, she cuts 2 inches off of one piece. How long is that piece now?

7. Mr. Cole bought 3 boxes of 10 markers each for his students to use. 4 of the markers are now missing. How many are left?

8. In 6 days, Jasmine biked 32 miles. If she biked 5 miles each of the first 5 days, how many miles did she bike on the 6th day?

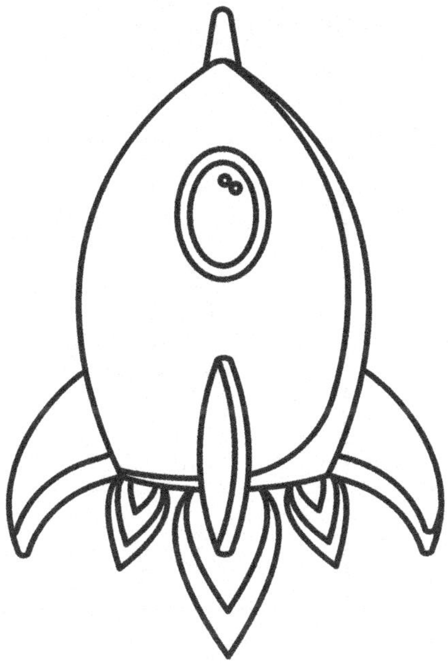

Fractions

Representing Fractions

Circle whether each shape is divided into equal or unequal shares.

	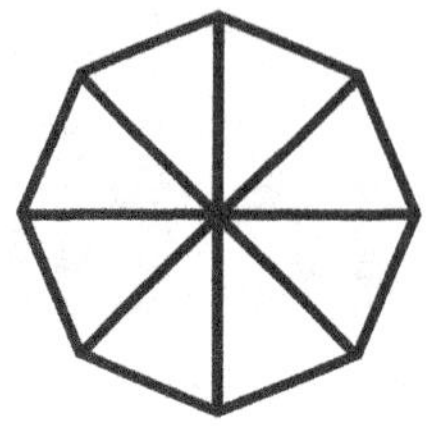	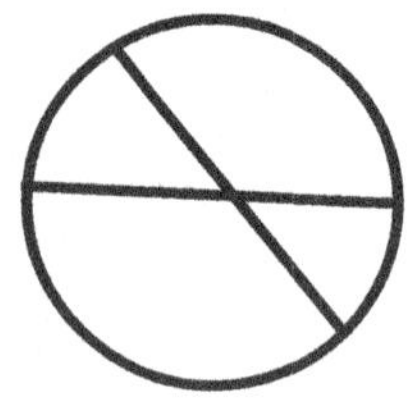	
equal	equal	equal	equal
unequal	unequal	unequal	unequal

Draw lines to divide each shape into equal shares.

1 

2 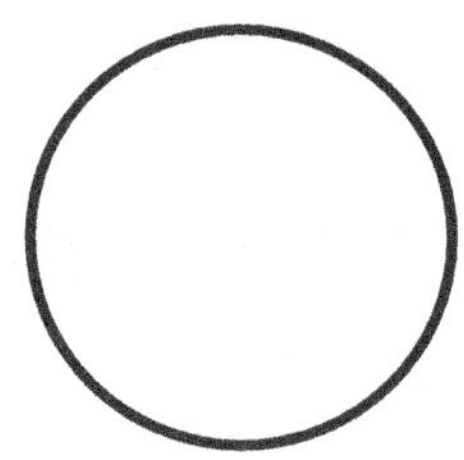

3 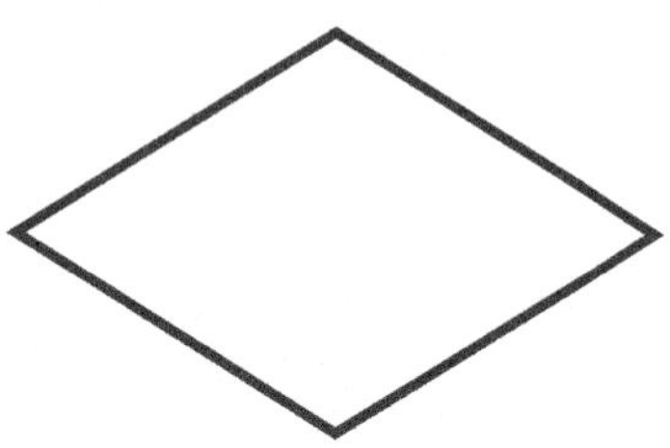

4 

5 

6 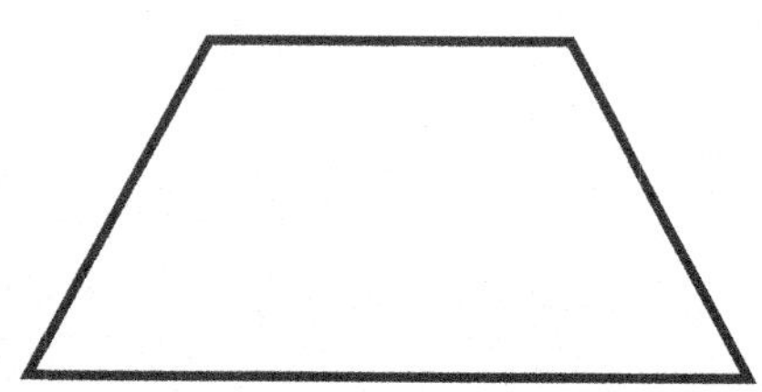

Write the number of equal shares shown in each shape. Then, write the name for the equal parts.

(EX)

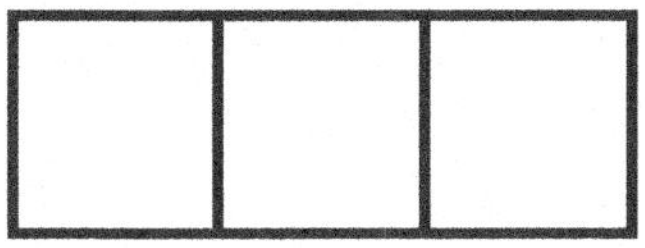

3 parts

name: thirds

(1)

_____ parts

name: __________

(2)

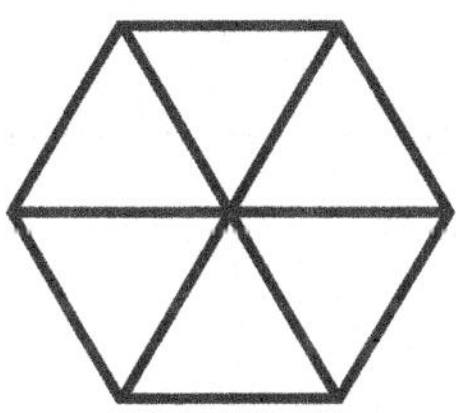

_____ parts

name: __________

(3)

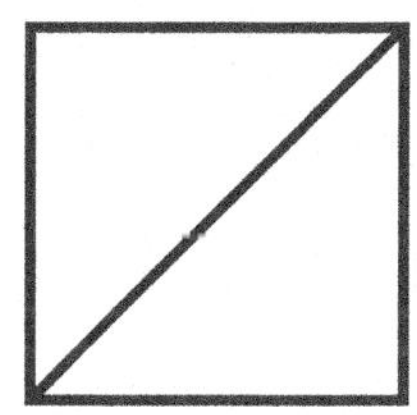

_____ parts

name: __________

(4)

_____ parts

name: __________

(5)

_____ parts

name: __________

Representing Fractions

Fill in the correct bubble to answer each question. Some questions may have more than one correct answer.

Which of these shapes are divided into fourths?

Which of these shapes are divided into thirds?

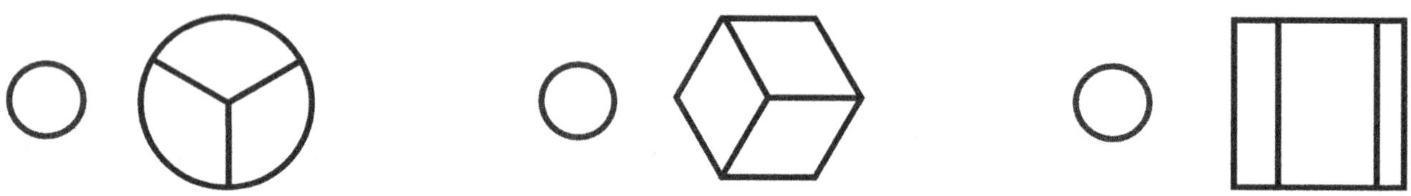

Which of these shapes are divided into eighths?

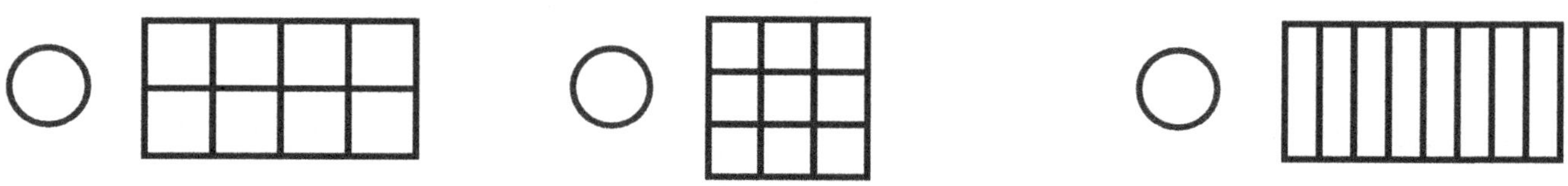

Which of these shapes are divided into halves?

Which of these shapes are divided into sixths?

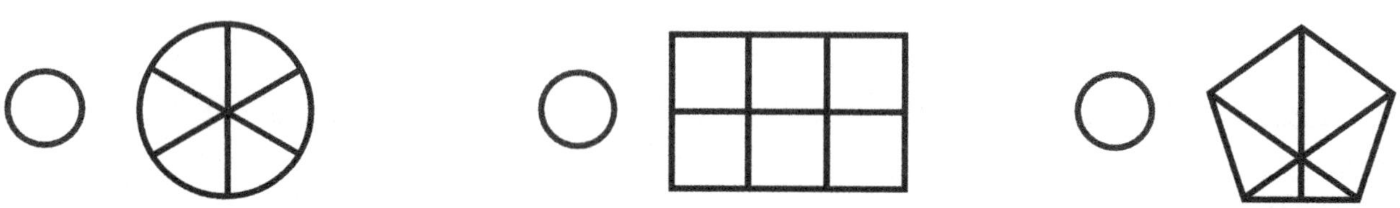

Representing Fractions

Write the fraction that names the shaded part of each whole.

EX 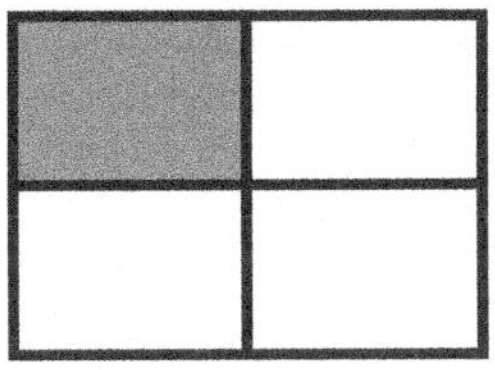 $\frac{1}{4}$

1 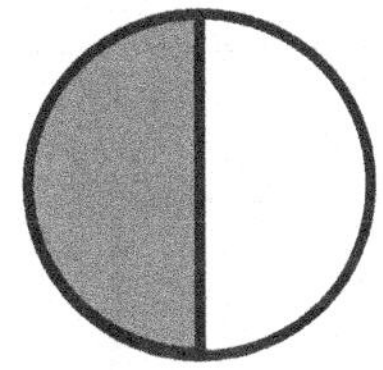 ____

2  ____

3  ____

4 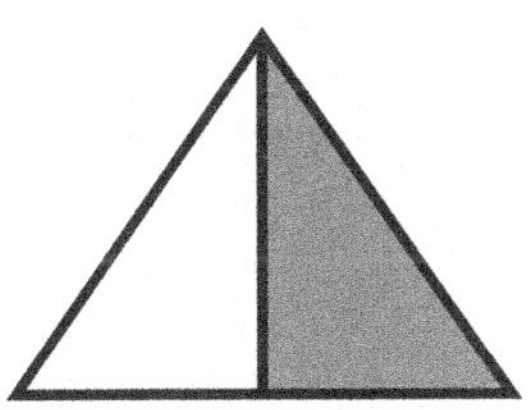 ____

5  ____

6 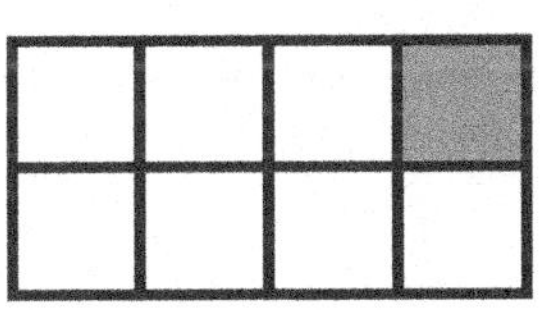 ____

7  ____

Representing Fractions

Divide each shape into equal shares and color to show the fraction.

$\frac{1}{4}$ 	$\frac{1}{2}$ 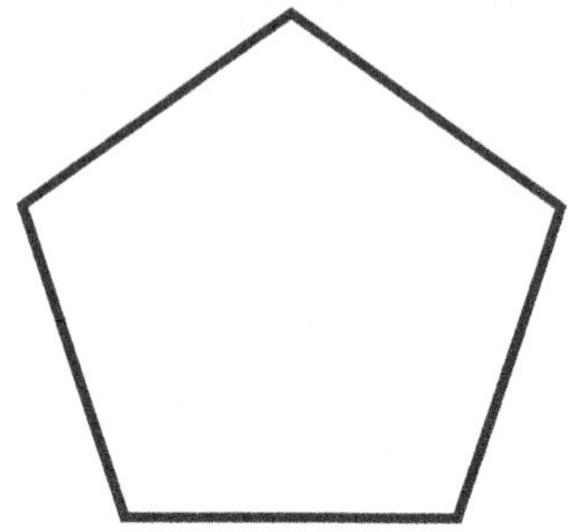
$\frac{1}{8}$ 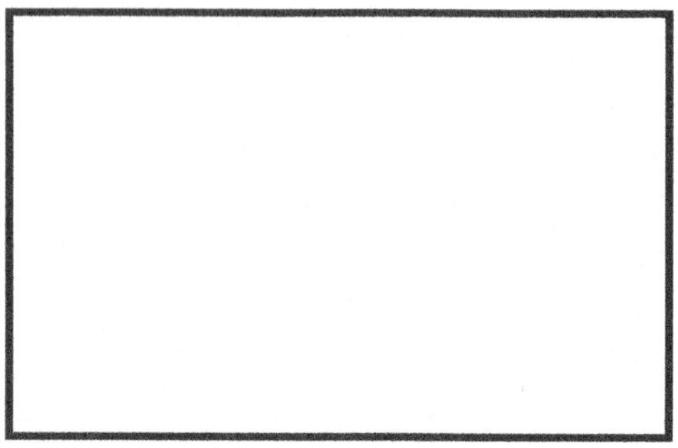	$\frac{1}{6}$ 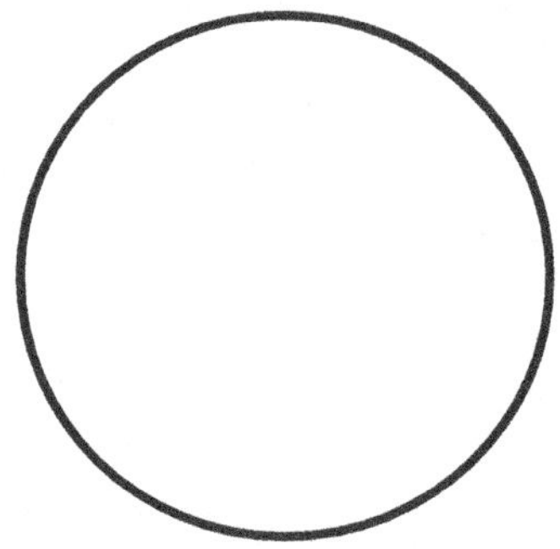
$\frac{1}{2}$ 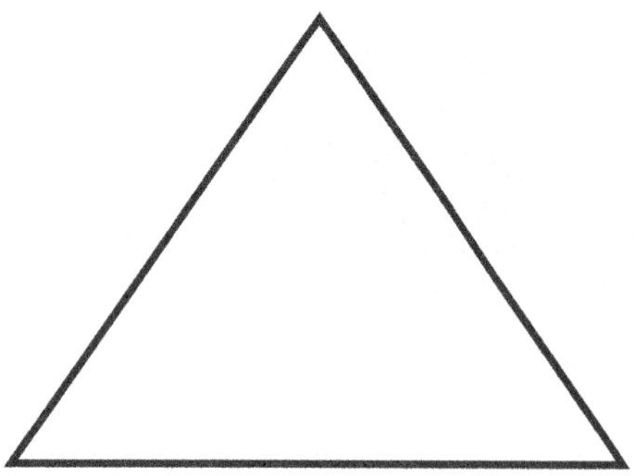	$\frac{1}{3}$ 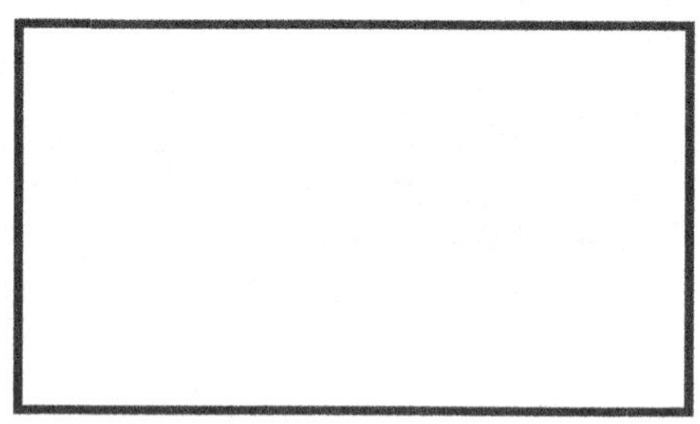
$\frac{1}{4}$ 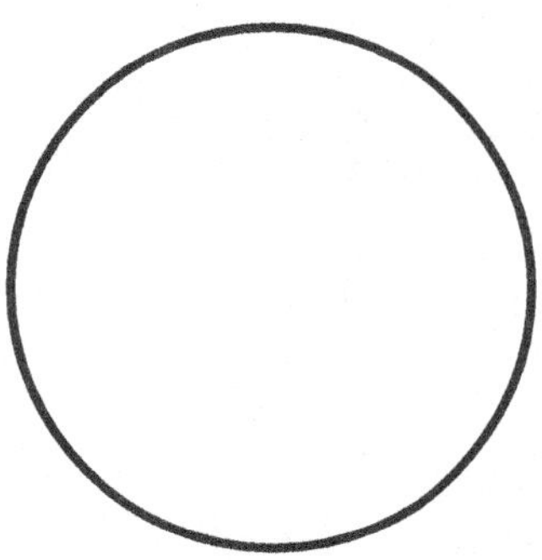	$\frac{1}{6}$ 

Representing Fractions

Color equal shares to show each fraction.

$\frac{1}{2}$ 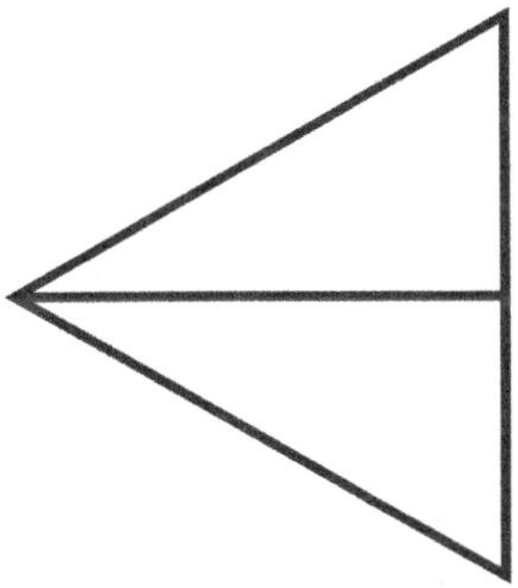	$\frac{4}{6}$ 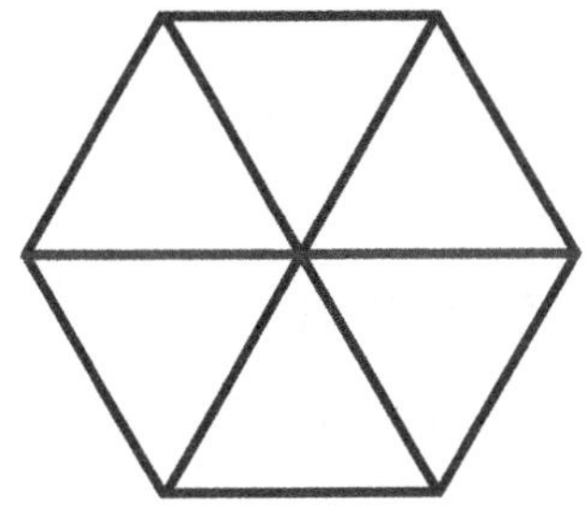
$\frac{3}{8}$ 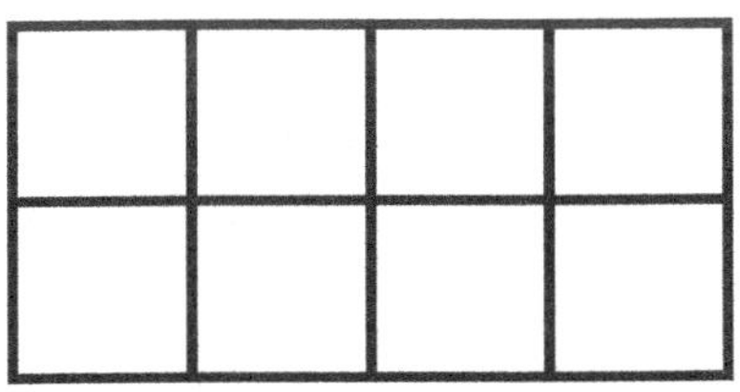	$\frac{2}{3}$ 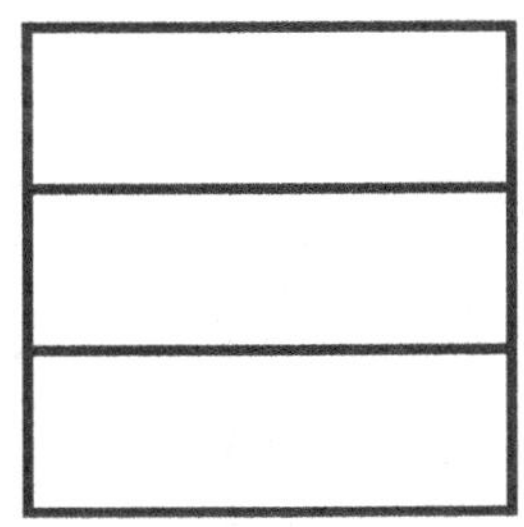
$\frac{3}{4}$ 	$\frac{2}{2}$ 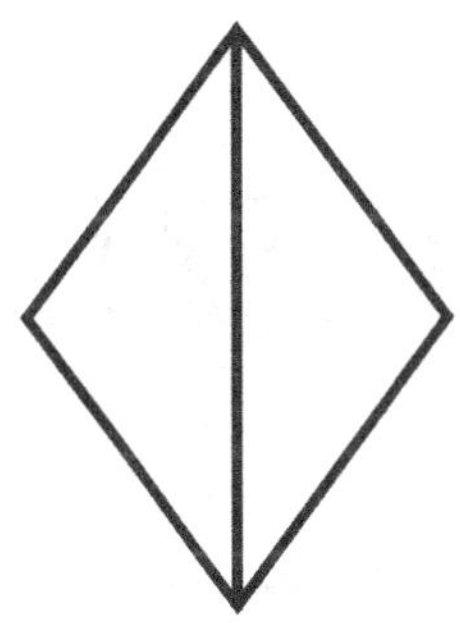
$\frac{5}{6}$ 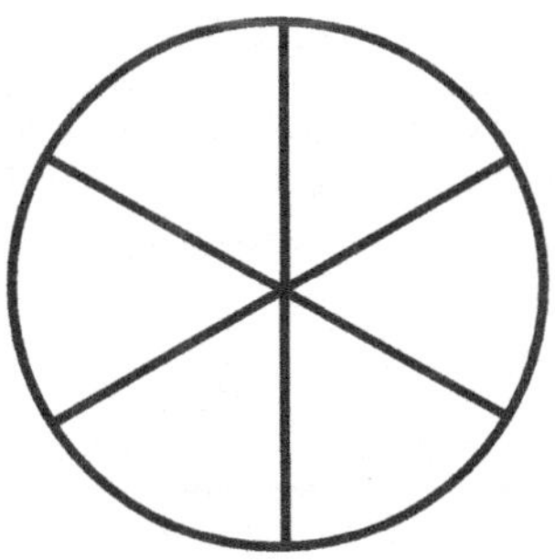	$\frac{2}{4}$ 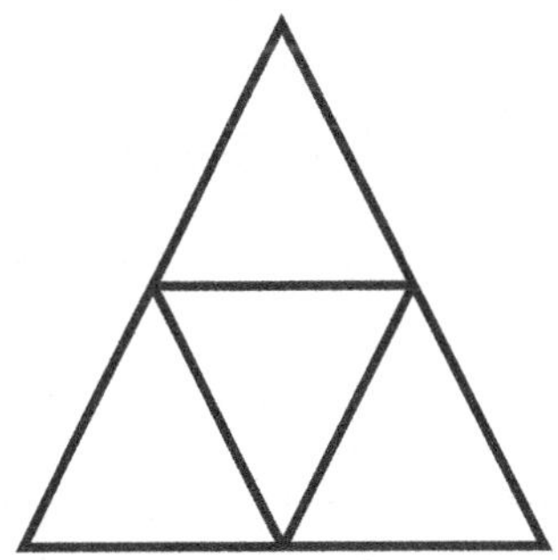

Representing Fractions

Fill in the bar model with the fraction that each share represents.

EX $\frac{1}{4}$ $\frac{1}{4}$ $\frac{1}{4}$ $\frac{1}{4}$

1

2

3

4

Color in shares of the bar model to show each fraction.

EX $\frac{3}{6}$

5 $\frac{7}{8}$

6 $\frac{2}{3}$

7 $\frac{3}{4}$

8 $\frac{1}{2}$

9 $\frac{5}{6}$

10 $\frac{3}{8}$

11 $\frac{2}{4}$

Representing Fractions

Divide each rectangle into a bar model to show each fraction.

EX $\frac{2}{6}$

1. $\frac{1}{3}$

2. $\frac{1}{4}$

3. $\frac{5}{8}$

4. $\frac{4}{6}$

5. $\frac{4}{4}$

6. $\frac{6}{8}$

7. $\frac{1}{2}$

Draw a bar model to represent each situation. Write the fraction on the line.

8. Greg has 8 books. He has read 4 of them. What fraction of his books has he read?

   ______

9. Isabelle has 3 cats. 2 of them are black. What fraction of her cats are NOT black?

   ______

10. Logan has 1 brother and 3 sisters. What fraction of his siblings are girls?

   ______

Representing Fractions

Label each number line with fractions.

EX
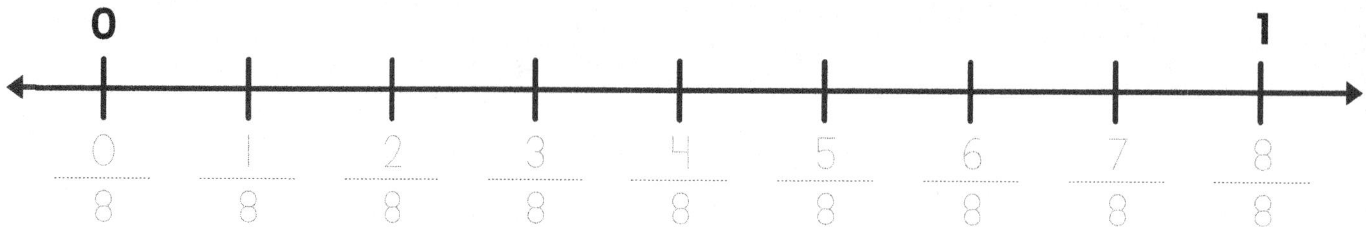

1
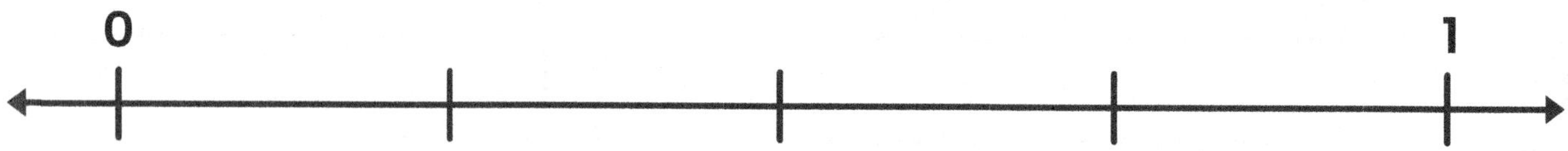

2
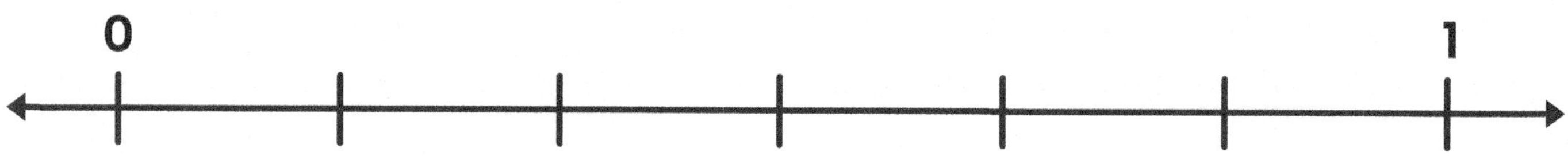

3

4
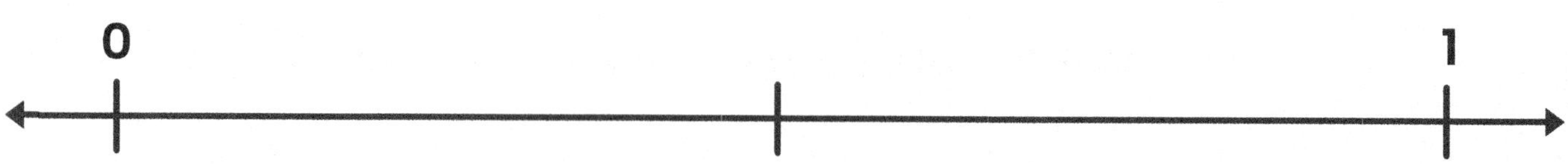

Representing Fractions

Write the fraction represented by the dot on each number line.

EX 

$\frac{2}{4}$

1 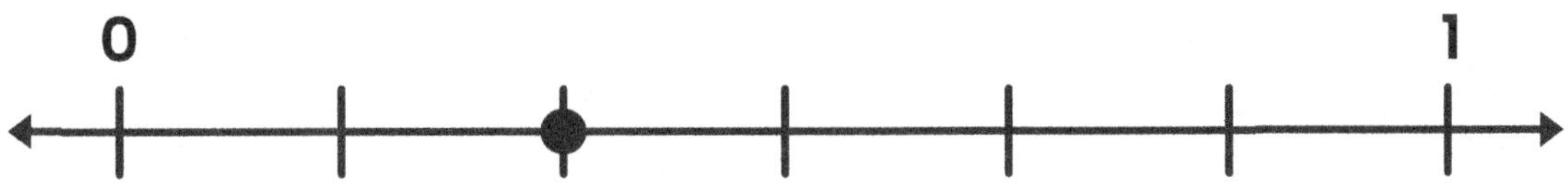

____

2 

____

3 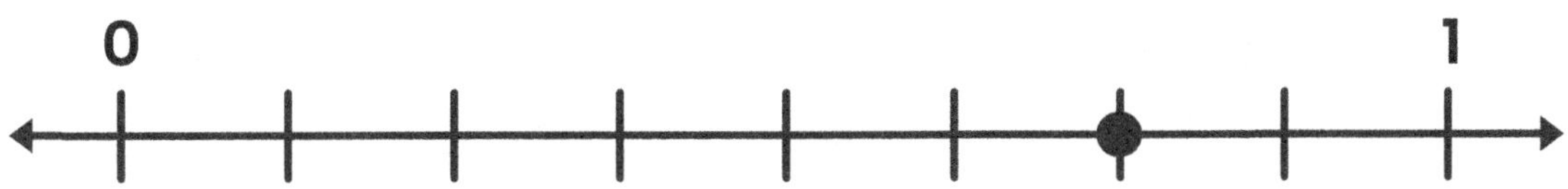

____

Draw a dot on each number line to represent the fraction shown.

4 $\frac{3}{8}$ 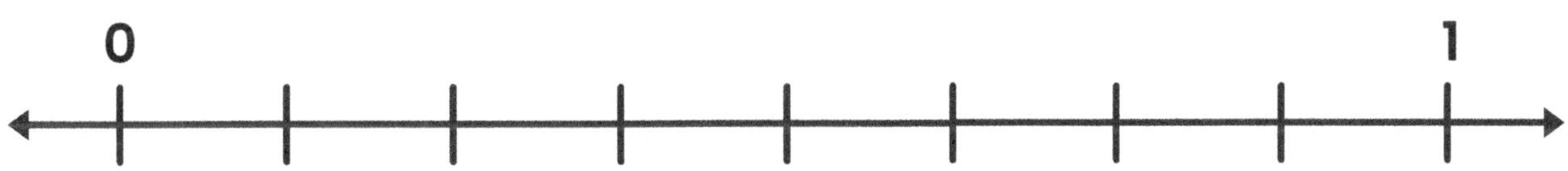

5 $\frac{1}{4}$ 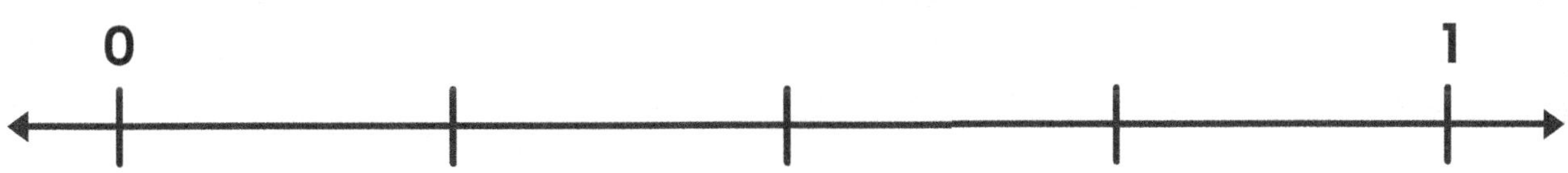

6 $\frac{4}{6}$ 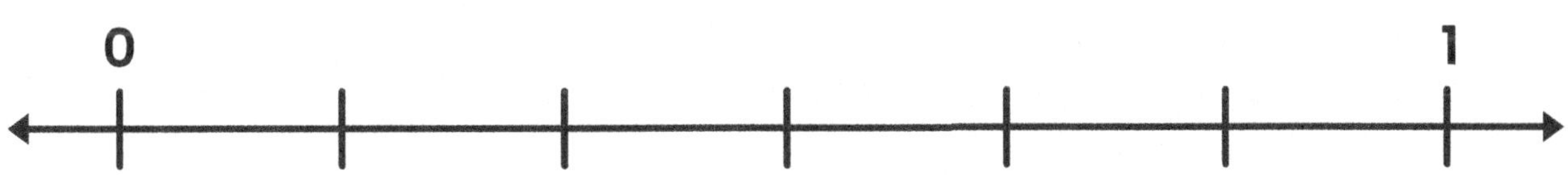

7 $\frac{2}{3}$ 

Representing Fractions

Draw any model to help you solve each word problem.

There are 5 boys and 3 girls in a group. What fraction of the group are girls?

____

A family of 6 is eating burgers for dinner. 4 of them have cheese on their burgers. What fraction of them do NOT have cheese on their burgers?

____

There are 4 photos hanging on the living room wall in Allie's house. Allie is in 3 of the photos. What fraction of the photos do NOT have Allie in them?

____

Equivalent Fractions

Color in shares on the bottom bar model to make a fraction equivalent to the one shown in the shaded portion of the top. Then, complete the equation.

EX

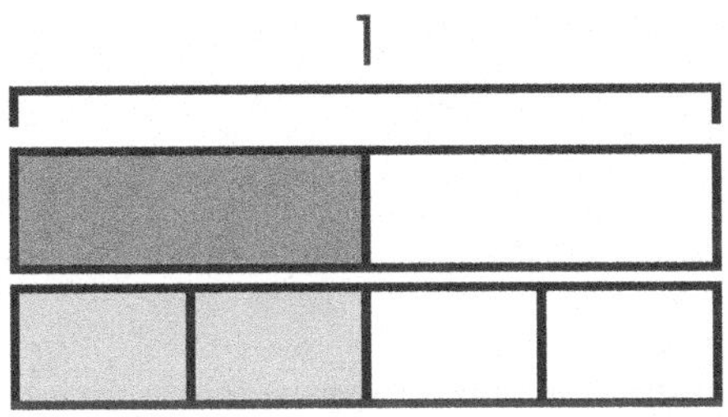

$$\frac{1}{2} = \frac{2}{4}$$

1

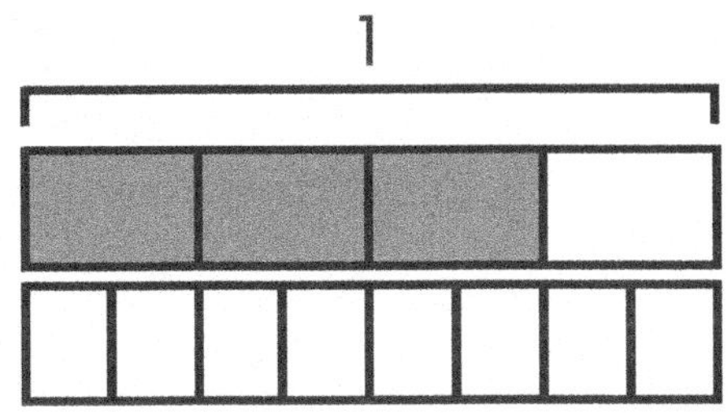

$$\frac{\square}{\square} = \frac{\square}{\square}$$

2

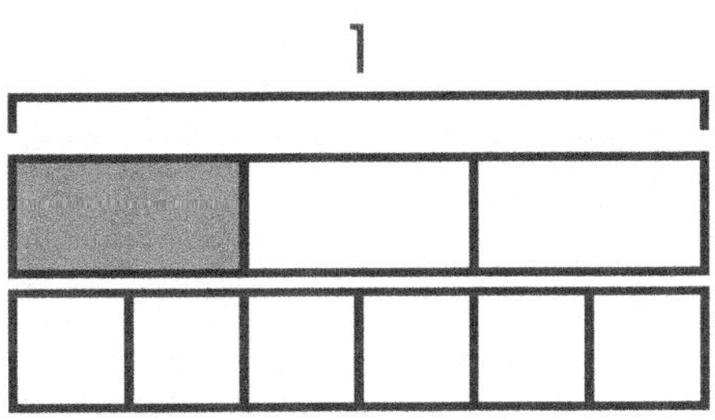

$$\frac{\square}{\square} = \frac{\square}{\square}$$

3

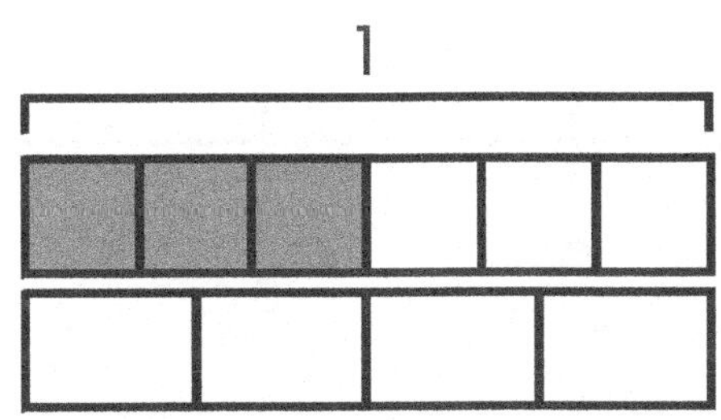

$$\frac{\square}{\square} = \frac{\square}{\square}$$

4

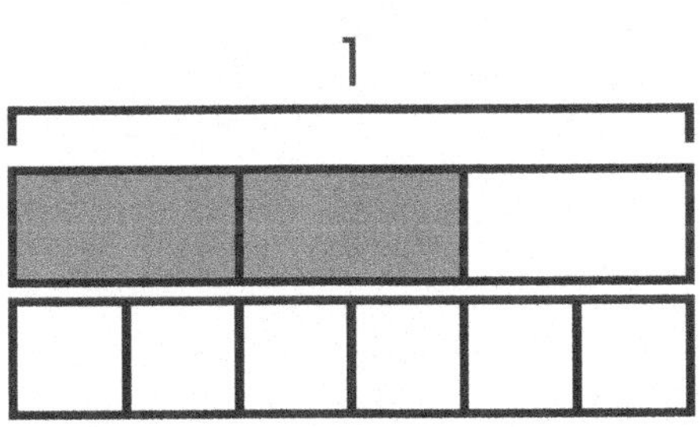

$$\frac{\square}{\square} = \frac{\square}{\square}$$

5

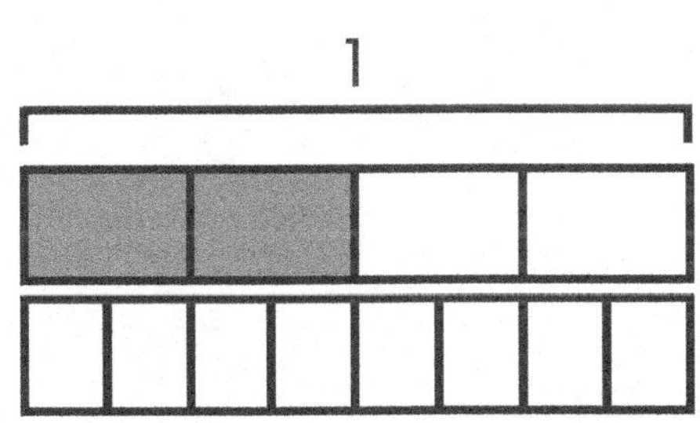

$$\frac{\square}{\square} = \frac{\square}{\square}$$

104 Equivalent Fractions

Use the number lines to find equivalent fractions.

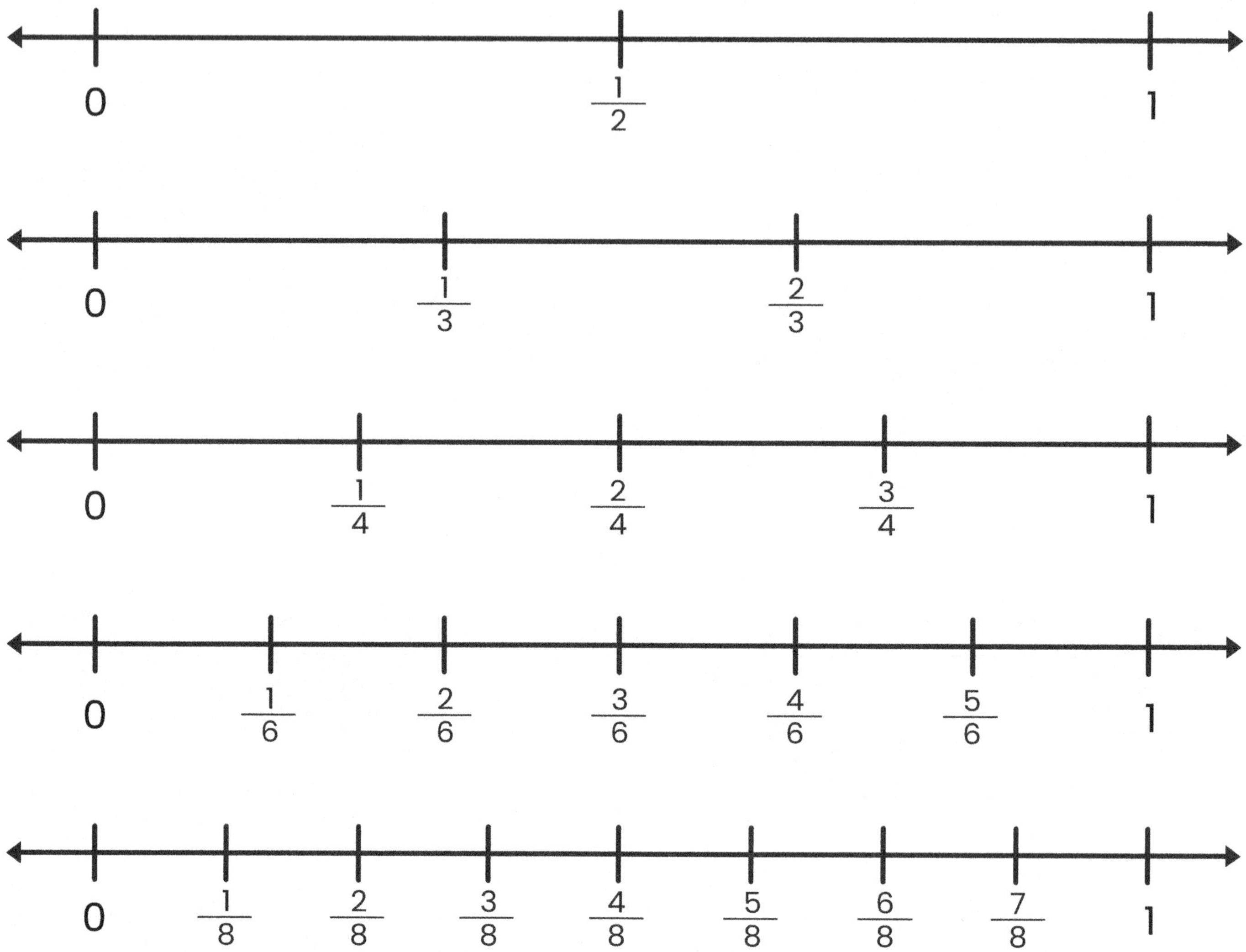

1. $\frac{1}{3} = \frac{\square}{6}$

2. $\frac{1}{2} = \frac{\square}{8}$

3. $\frac{6}{8} = \frac{\square}{4}$

4. $\frac{4}{6} = \frac{\square}{3}$

5. $\frac{2}{4} = \frac{\square}{8}$

6. $\frac{1}{2} = \frac{\square}{6}$

Equivalent Fractions

Graph each number on the number line.

EX $\frac{1}{4}$

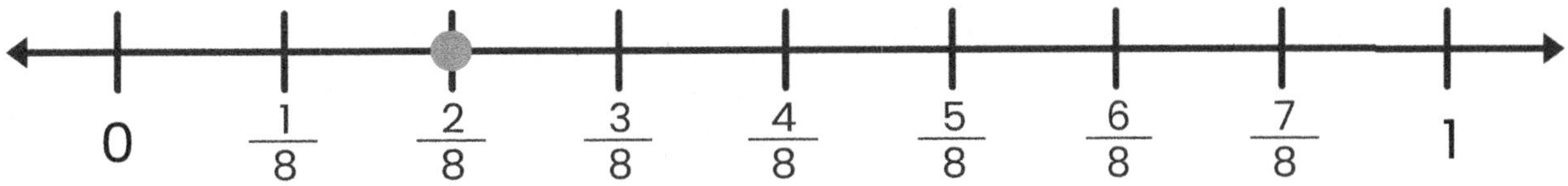

1 $\frac{3}{4}$

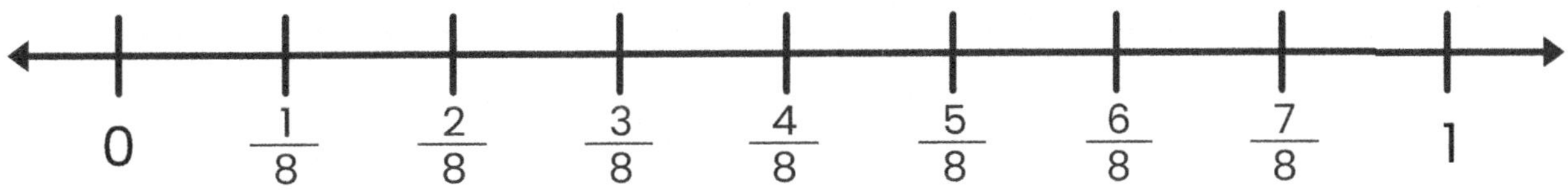

2 $\frac{1}{2}$

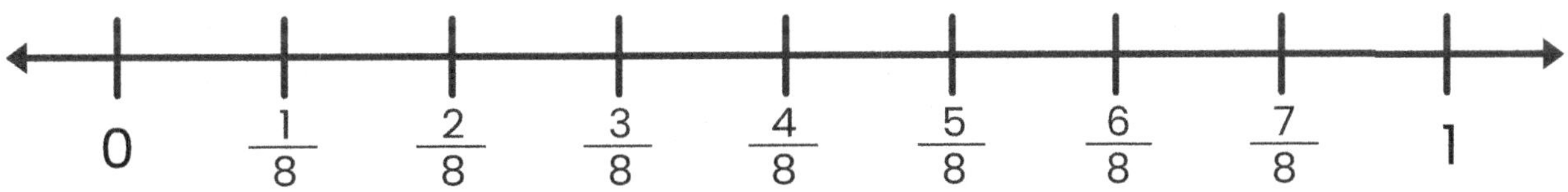

3 $\frac{1}{3}$

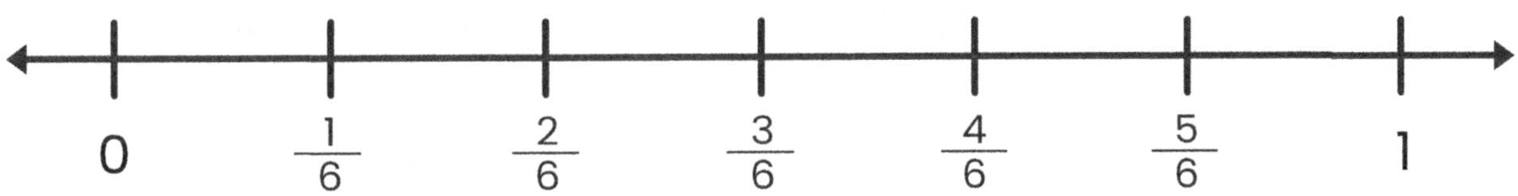

4 $\frac{1}{2}$

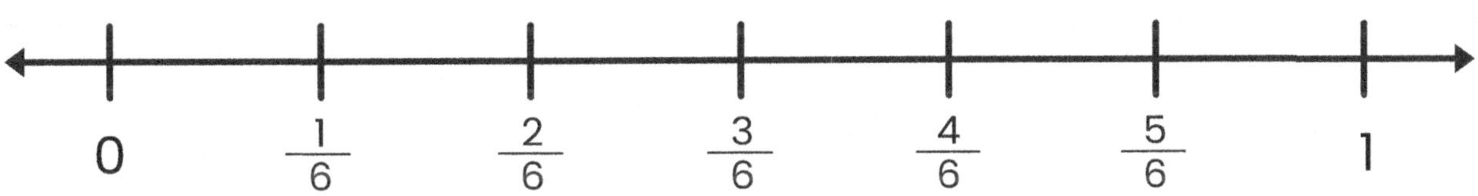

5 $\frac{2}{3}$

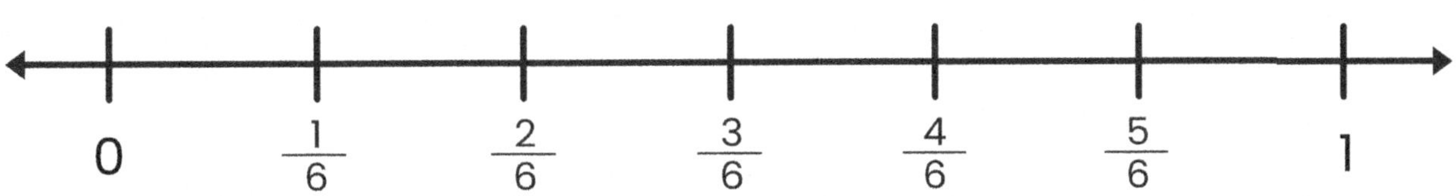

Equivalent Fractions

Circle the fractions in each row that are equivalent to the first fraction. Put a star on the fraction in each row that is in lowest terms.

$\frac{4}{8}$	$\frac{2}{4}$	$\frac{1}{3}$	$\frac{1}{2}$	$\frac{5}{6}$
$\frac{4}{6}$	$\frac{3}{4}$	$\frac{2}{3}$	$\frac{6}{8}$	$\frac{1}{3}$
$\frac{2}{4}$	$\frac{4}{6}$	$\frac{1}{2}$	$\frac{3}{6}$	$\frac{2}{3}$
$\frac{1}{3}$	$\frac{2}{6}$	$\frac{4}{8}$	$\frac{3}{6}$	$\frac{2}{4}$
$\frac{3}{4}$	$\frac{1}{2}$	$\frac{3}{8}$	$\frac{4}{6}$	$\frac{6}{8}$
$\frac{1}{2}$	$\frac{3}{6}$	$\frac{2}{4}$	$\frac{3}{8}$	$\frac{1}{3}$

Fill in the correct bubble to answer each question.

Which of these is equal to 3?

○ $\frac{3}{6}$ ○ $\frac{3}{3}$ ○ $\frac{3}{1}$

Which of these is equal to 1?

○ $\frac{6}{1}$ ○ $\frac{1}{2}$ ○ $\frac{8}{8}$

Which of these is equal to 4?

○ $\frac{4}{1}$ ○ $\frac{2}{8}$ ○ $\frac{4}{4}$

Which of these is NOT equal to 1?

○ $\frac{2}{2}$ ○ $\frac{1}{1}$ ○ $\frac{8}{1}$

Which of these is equal to $\frac{6}{6}$?

○ $\frac{5}{5}$ ○ $\frac{3}{1}$ ○ $\frac{2}{3}$

108 Comparing Fractions

Write the fraction shown by the shaded region of each model. Then, use >, <, or = to compare each pair of fractions.

EX

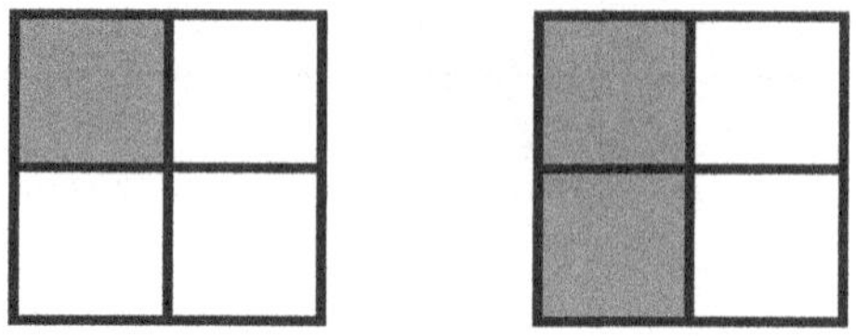

$\frac{1}{4}$ < $\frac{2}{4}$

1

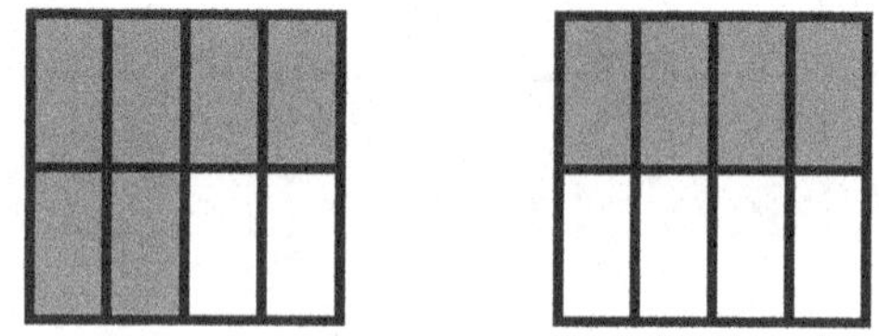

_____ ◯ _____

2

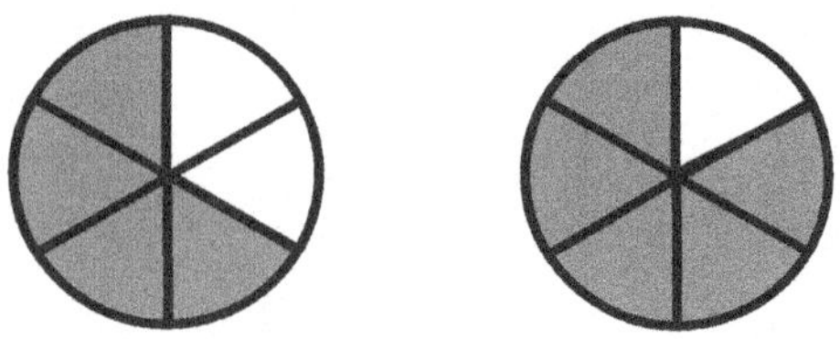

_____ ◯ _____

3

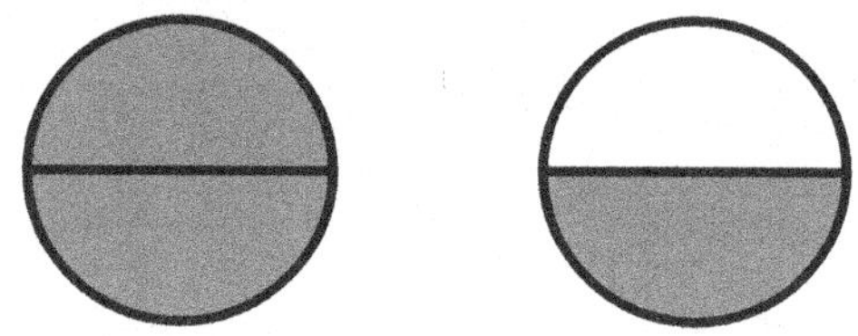

_____ ◯ _____

4

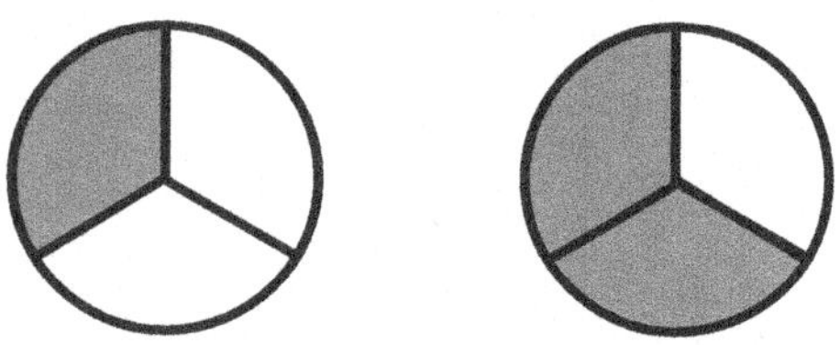

_____ ◯ _____

5

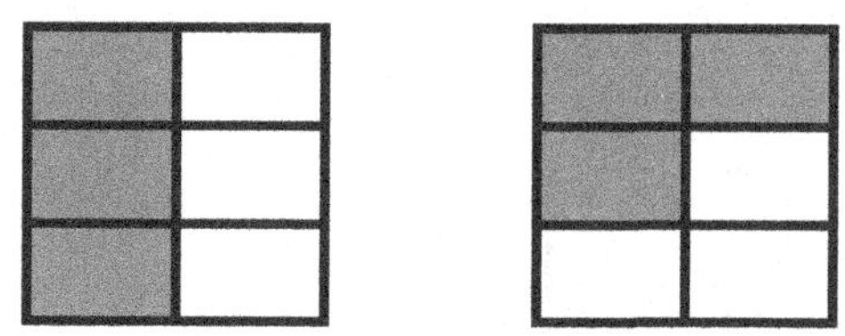

_____ ◯ _____

Use >, <, or = to compare each pair of fractions.

EX $\frac{1}{3}$ < $\frac{2}{3}$

6 $\frac{3}{4}$ ◯ $\frac{2}{4}$

7 $\frac{1}{2}$ ◯ $\frac{1}{2}$

8 $\frac{4}{8}$ ◯ $\frac{5}{8}$

9 $\frac{2}{6}$ ◯ $\frac{1}{6}$

10 $\frac{3}{3}$ ◯ $\frac{2}{3}$

Comparing Fractions

Write the fraction shown by the shaded region of each model. Then, use >, <, or = to compare each pair of fractions.

(EX)

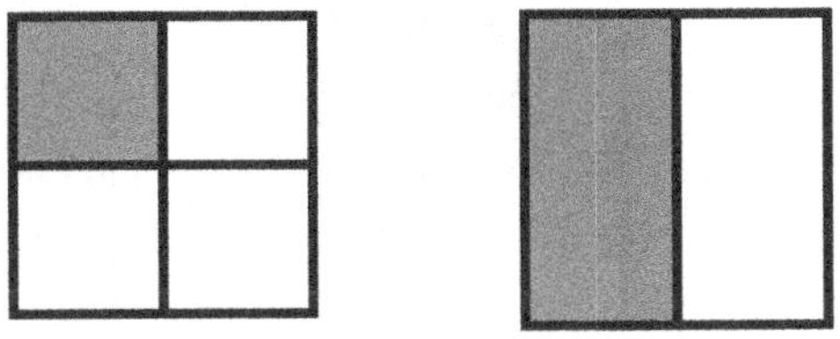

$\frac{1}{4} < \frac{1}{2}$

(1)

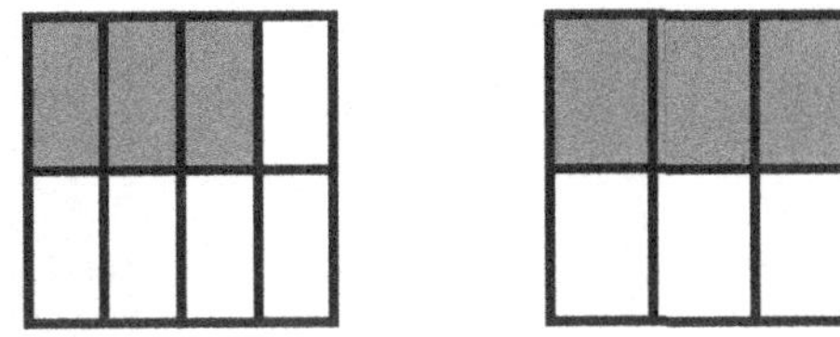

______ ◯ ______

(2)

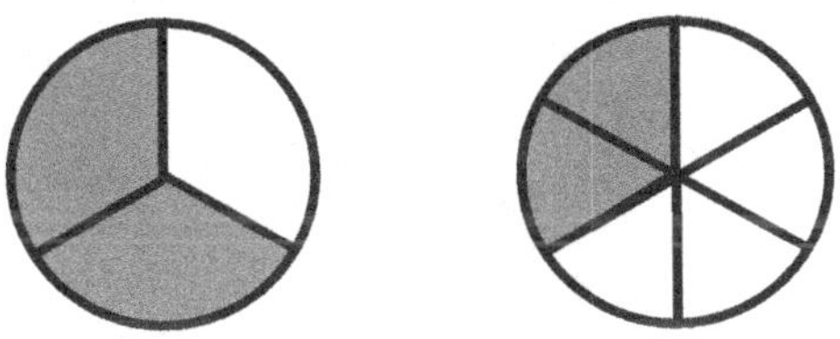

______ ◯ ______

(3)

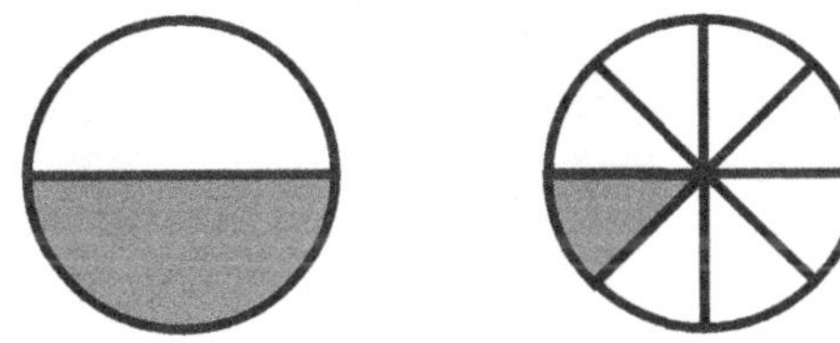

______ ◯ ______

(4)

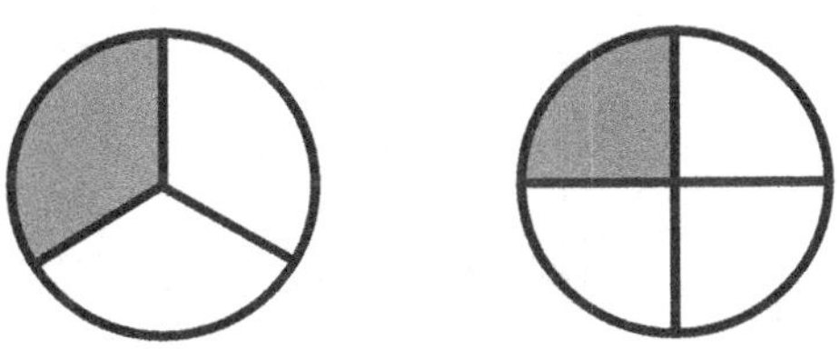

______ ◯ ______

(5)

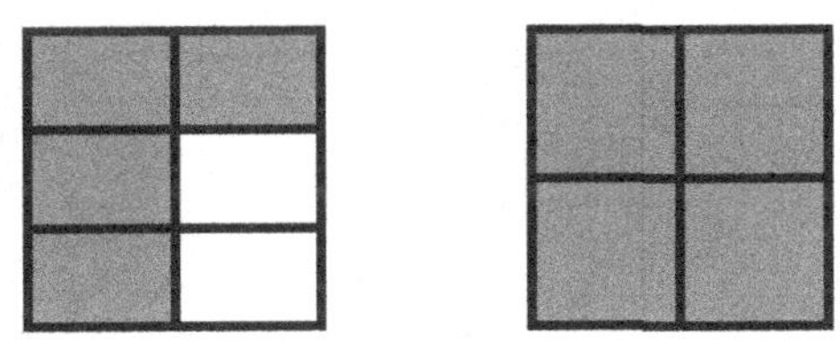

______ ◯ ______

Use >, <, or = to compare each pair of fractions.

(EX) $\frac{1}{3} < \frac{1}{2}$

(6) $\frac{1}{6}$ ◯ $\frac{1}{8}$

(7) $\frac{2}{4}$ ◯ $\frac{2}{6}$

(8) $\frac{2}{8}$ ◯ $\frac{2}{3}$

(9) $\frac{3}{6}$ ◯ $\frac{3}{3}$

(10) $\frac{4}{6}$ ◯ $\frac{4}{8}$

Comparing Fractions

Write the fraction shown by the shaded region of each model. Then, use >, <, or = to compare each pair of fractions.

EX

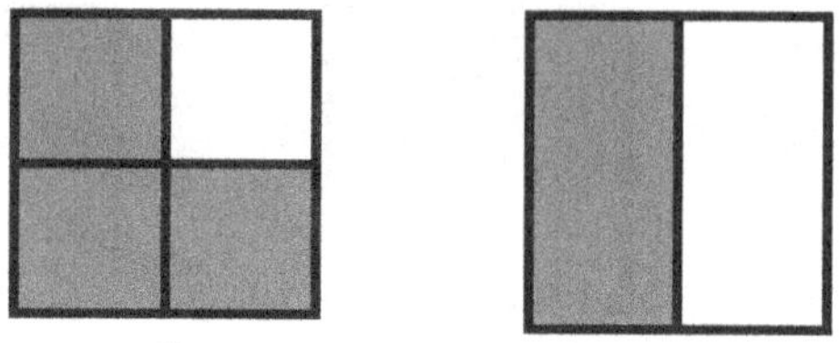

$\frac{3}{4}$ > $\frac{1}{2}$

1

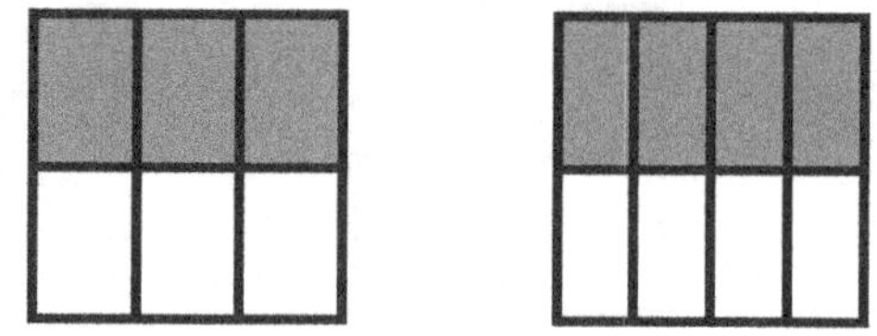

______ ◯ ______

2

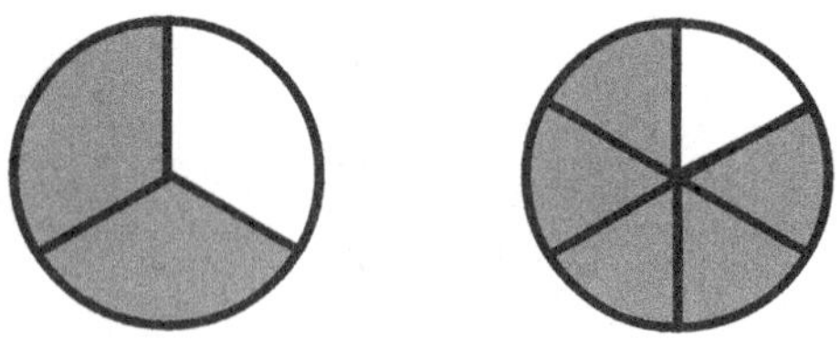

______ ◯ ______

3

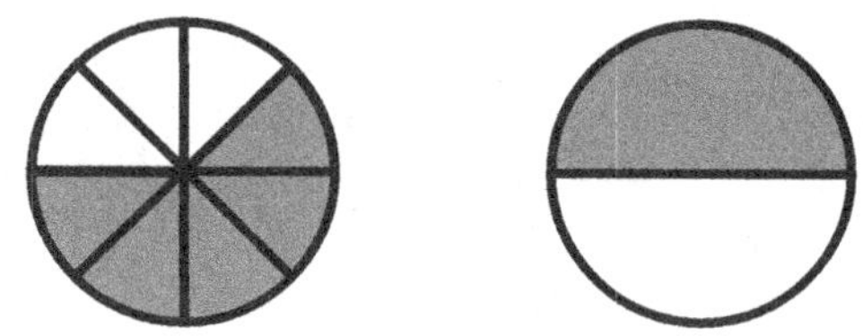

______ ◯ ______

4

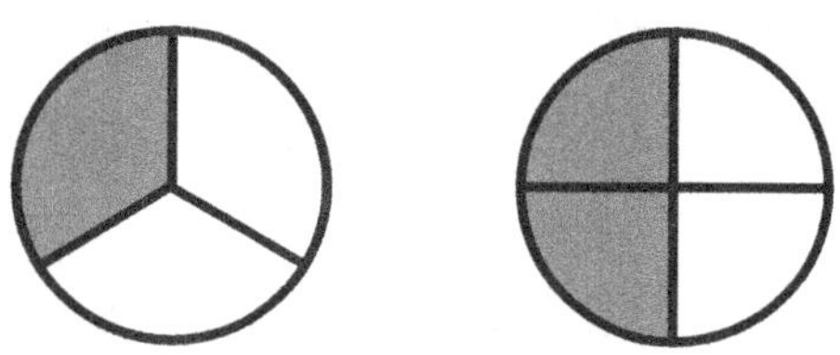

______ ◯ ______

5

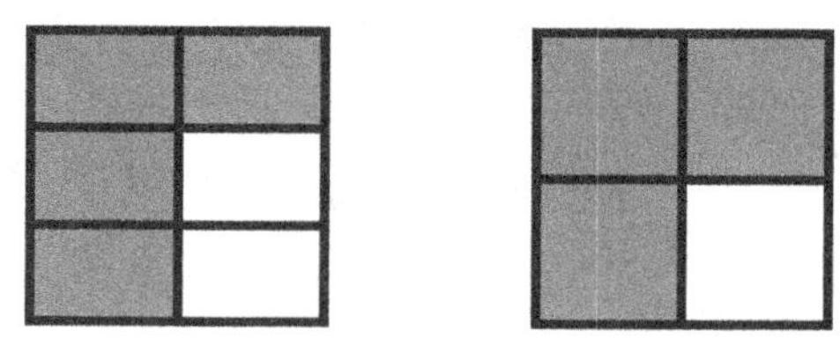

______ ◯ ______

Use >, <, or = to compare each pair of fractions.

EX $\frac{1}{3}$ < $\frac{3}{4}$

6 $\frac{3}{6}$ ◯ $\frac{2}{3}$

7 $\frac{3}{8}$ ◯ $\frac{1}{2}$

8 $\frac{2}{8}$ ◯ $\frac{1}{4}$

9 $\frac{5}{6}$ ◯ $\frac{3}{4}$

10 $\frac{2}{3}$ ◯ $\frac{7}{8}$

Comparing Fractions

Graph each pair of fractions on the number line. Then, circle the larger fraction.

EX $\frac{1}{4}$ $\frac{3}{8}$

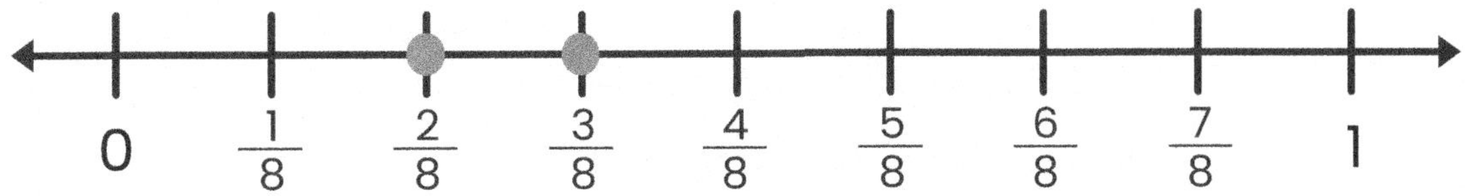

1 $\frac{3}{4}$ $\frac{5}{8}$

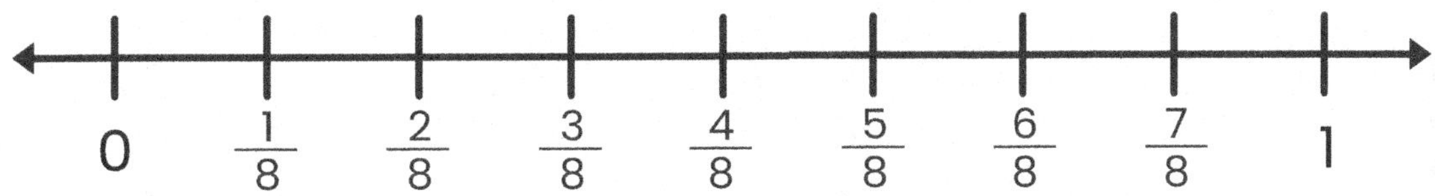

2 $\frac{1}{2}$ $\frac{6}{8}$

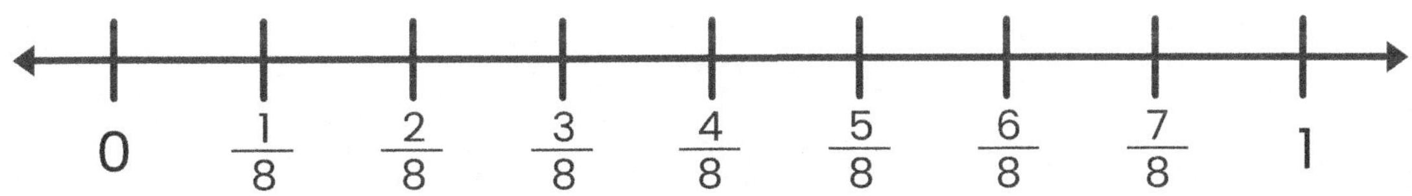

3 $\frac{1}{3}$ $\frac{1}{6}$

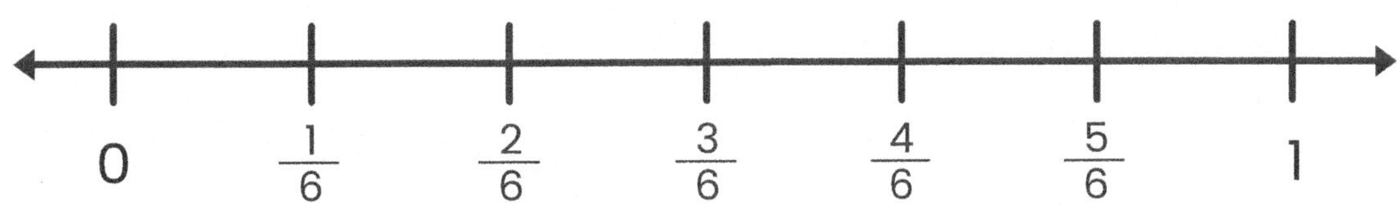

4 $\frac{2}{3}$ $\frac{5}{6}$

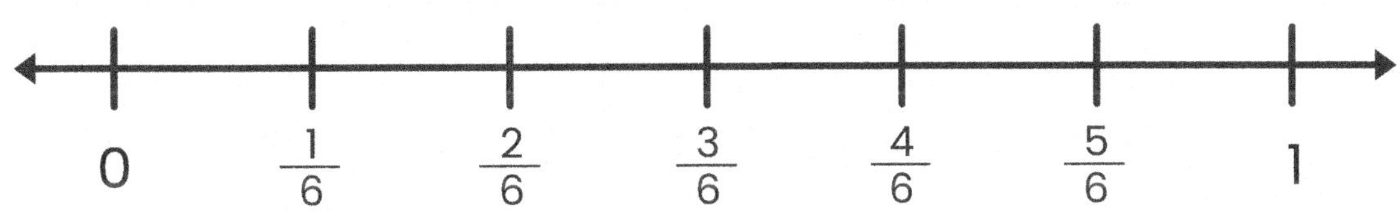

5 $\frac{4}{6}$ $\frac{1}{2}$

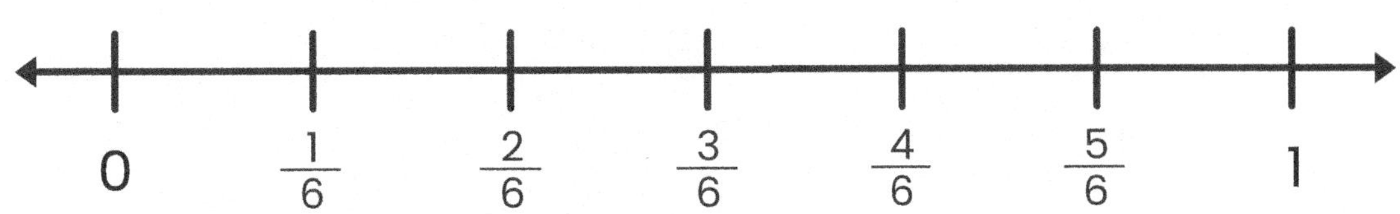

Comparing Fractions

Draw any visual model to help you answer each word problem.

Nina practiced violin for $\frac{1}{2}$ hour on Monday, for $\frac{3}{4}$ hour on Tuesday, and for $\frac{2}{3}$ hour on Wednesday. On which of the three days she she practice the longest?

Kaya is frosting 24 cupcakes. She puts pink frosting on $\frac{1}{2}$ of the cupcakes, blue on $\frac{1}{3}$ of them, and green on $\frac{1}{6}$ of them. Which color does she use on the most cupcakes?

Alicia and Reagan are working on the same homework assignment. Alicia has finished $\frac{5}{6}$ of the problems so far, and Reagan has finished $\frac{2}{3}$. Who has finished more of the assignment?

Fill in the correct bubble to answer each question.

A recipe calls for 1 cup of flour. Kyle has a $\frac{1}{3}$ cup scoop. How many scoops of flour will he need?

- ○ 1
- ○ 2
- ○ 3

A recipe calls for $\frac{1}{2}$ teaspoon of salt, $\frac{1}{4}$ teaspoon of pepper, and $\frac{2}{3}$ teaspoon of cumin. Which ingredient is the least amount?

- ○ salt
- ○ pepper
- ○ cumin

Damian is cutting a pizza into equal shares. Which of these shares would be the largest slices of pizza?

- ○ fourths
- ○ sixths
- ○ eighths

Lena, Nicholas, and Mark count the vowels in their names and write a fraction showing the number of vowels out of the total number of letters in each name. Whose fraction is the largest?

- ○ Lena
- ○ Nicholas
- ○ Mark

Micah has a measuring cup that holds a fourth of a cup and one that holds half of a cup. How many fourth-cup scoops would equal 2 half-cups?

- ○ 2
- ○ 4
- ○ 6

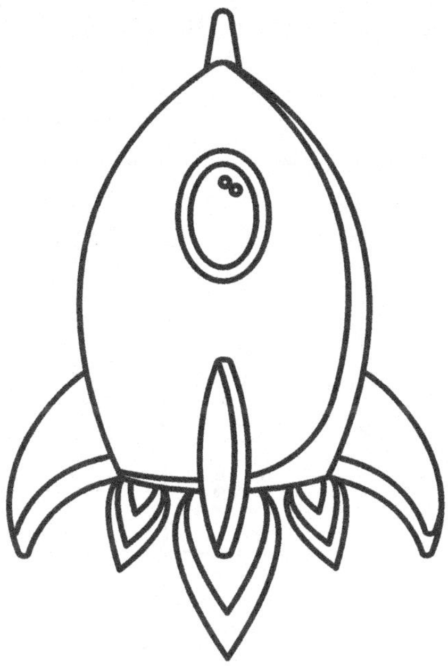

Measurement and Data

Use the ruler shown to measure the length of each object to the nearest $\frac{1}{2}$ inch.

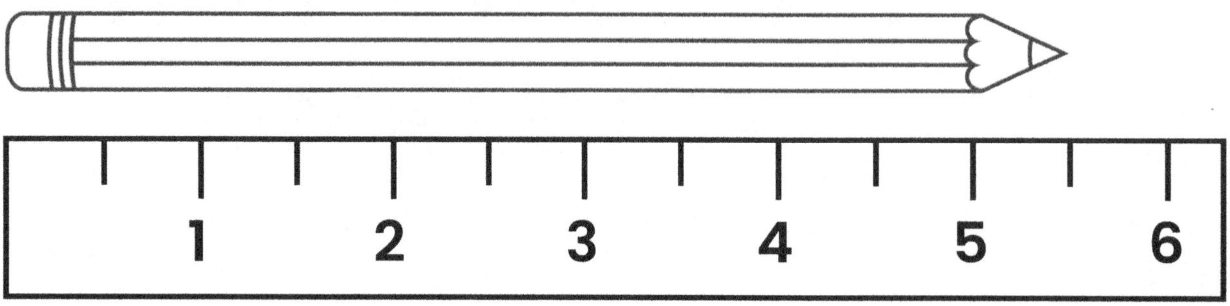

______ inches

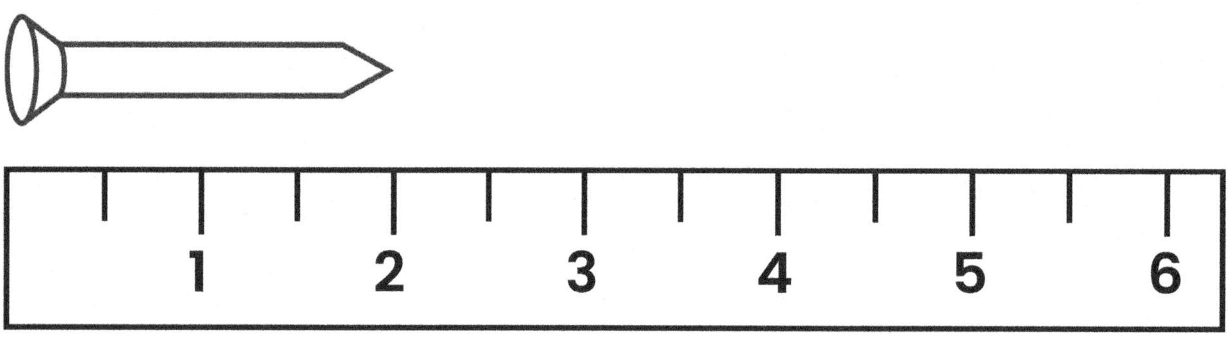

______ inches

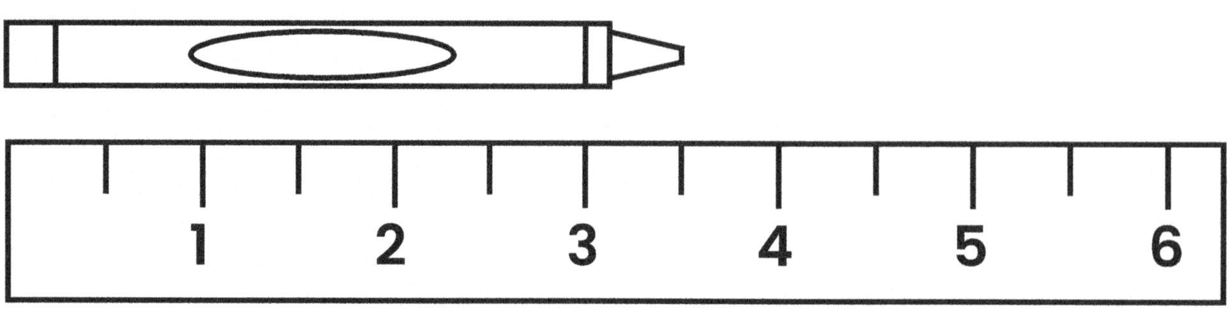

______ inches

Use the ruler shown to measure the length of each object to the nearest $\frac{1}{4}$ inch.

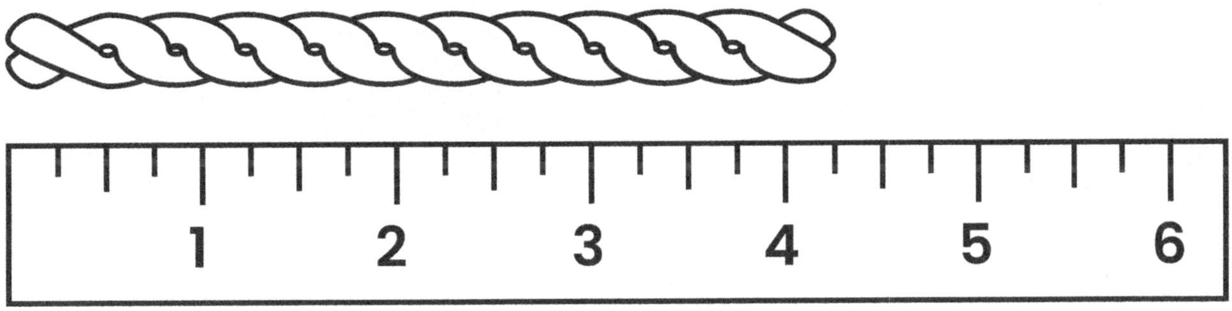

_______ inches

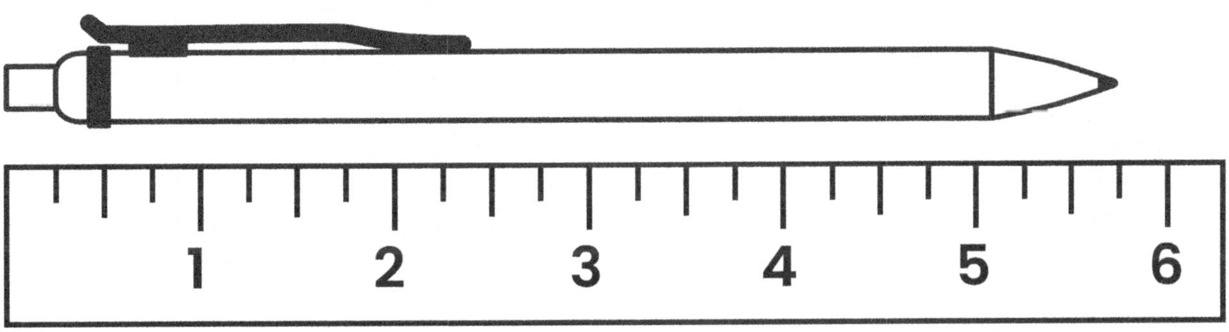

_______ inches

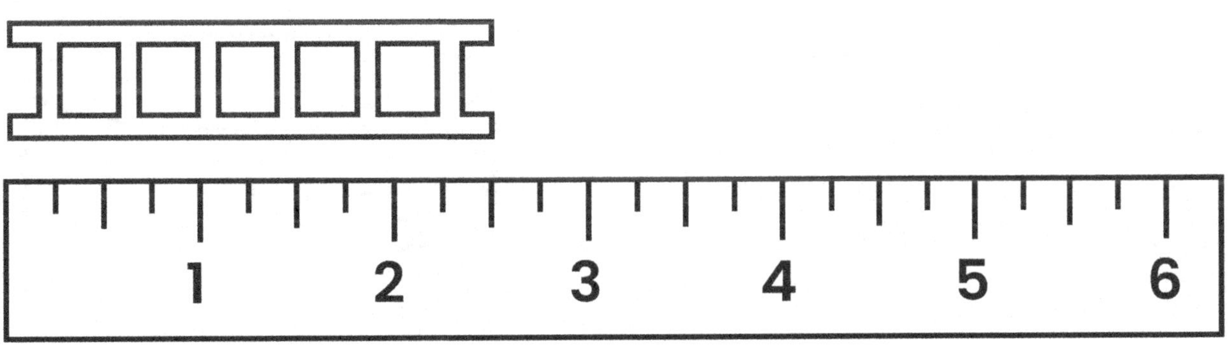

_______ inches

Find the volume of liquid in each container shown.

1

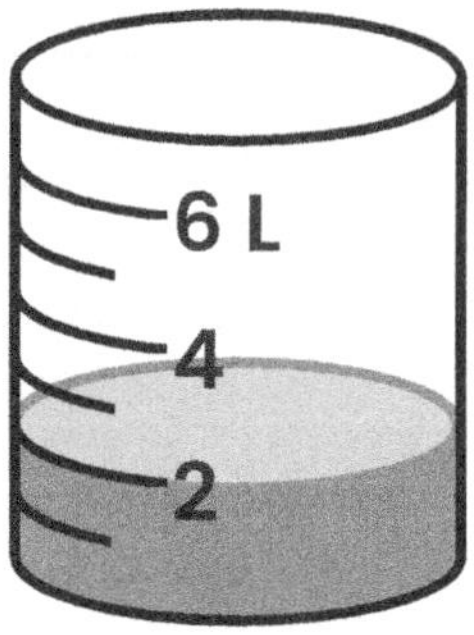

______ liters

2

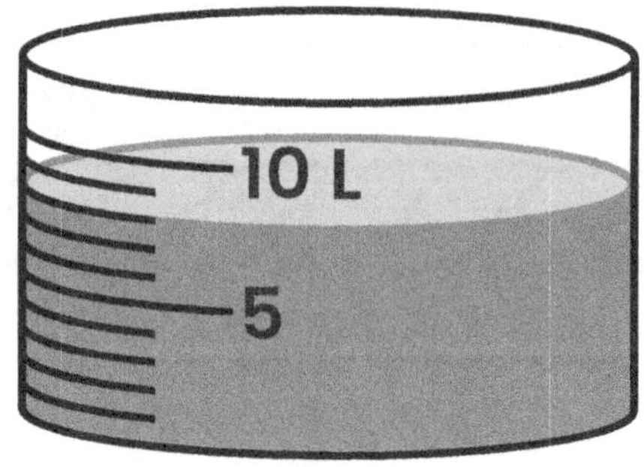

______ liters

3

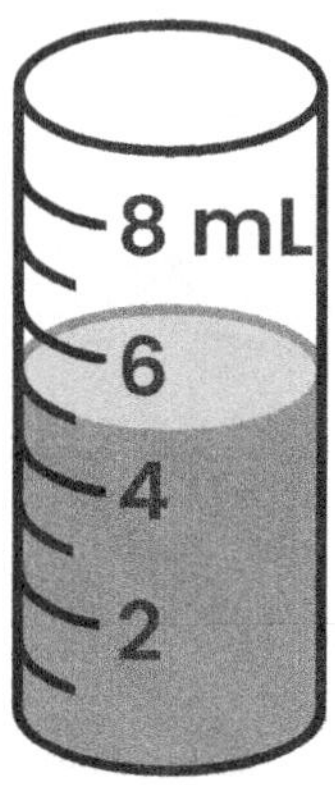

______ milliliters

4

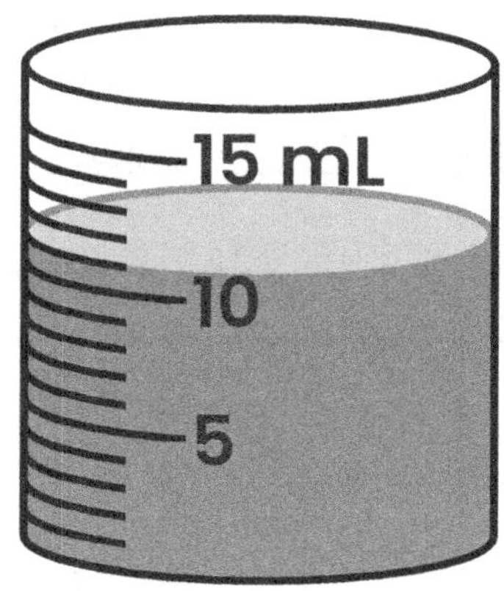

______ milliliters

---

5 Drew has 7 liters of water in a container. He pours out 3 liters. How many liters are left? Draw to show the remaining water in the container.

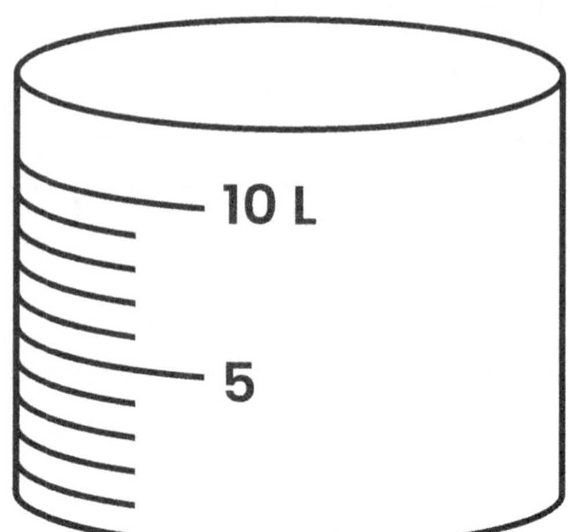

______ liters

Answer each question about mass based on the pictures shown.

1. The apple has a mass of 100 grams. The difference in mass between the apple and the lemon is 15 grams. What is the mass of the lemon?

______ grams

2. Three different weights, labeled A, B, and C, are placed on the scale. Which has the greatest mass? Explain how you know.

____

____

____

3. The masses (in grams) of two weights are shown in the picture. What is the mass of the weight marked X?

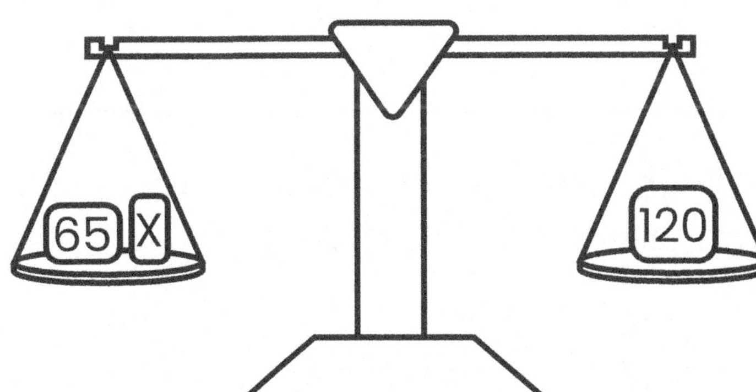

______ grams

Measurement

Fill in the correct bubble to answer each question.

Which of these most likely contains a volume of liquid close to 1 liter?

- ◯ A bottle of eye drops
- ◯ A kiddie pool
- ◯ A bottle of shampoo

Which of these is NOT a measure of mass?

- ◯ liter
- ◯ gram
- ◯ kilogram

Which of these is most likely to be the mass of a car?

- ◯ 10 kg
- ◯ 1000 g
- ◯ 1000 kg

Which of these is most likely to be the volume of liquid in a full juice box?

- ◯ 10 mL
- ◯ 120 mL
- ◯ 2 L

Which of these masses would you most likely measure in grams rather than kilograms?

- ◯ A horse
- ◯ A cell phone
- ◯ A bowling ball

Solve each word problem about volume or mass.

1. Container A contains 6 L of water. Container B contains 12 L of water. Gina pours water from Container B into Container A until they contain the same volume of water. How much water is in each container now?

   ______ liters

2. Ryan measures the mass of his three dogs. Fido's mass is 23 kg, Spot's mass is 18 kg, and Rover's mass is 29 kg. What is the difference in mass between the heaviest and lightest dogs?

   ______ kg

3. Sheila has 40 identical marbles, each with a mass of 3 grams. What is the total mass of the marbles?

   ______ grams

4. A beaker contains 80 mL of liquid. Ruby uses a dropper that holds 5 mL to remove liquid. If she fills her dropper 6 times with liquid from the beaker, how much liquid is left in the beaker when she finishes?

   ______ mL

Representing Data

Use the information and picture graph shown below to answer questions 1-4.

A farmer recorded how many of each type of apple he sold at his stand on a certain day. The results are shown in the picture graph.

1. How many of each types of apple did he sell?

   Gala: ______ Granny Smith: ______ Fuji: ______

2. How many apples were sold all together?

3. How many more Galas did he sell than Fujis?

4. How many fewer Granny Smiths did he sell than Fujis?

Use the information below to create a picture graph. Then, answer the questions that follow.

A library tracked the number of novels that were borrowed by genre. On an average day, library patrons borrowed 45 realistic fiction novels, 35 fantasy novels, 25 historical fiction novels, 10 science fiction novels, 15 mystery novels, and 15 thrillers.

Create a picture graph to represent this data. Choose a symbol to represent a quantity of books greater than 1.

NOVELS BORROWED PER DAY	
Realistic Fiction	
Fantasy	
Historical Fiction	
Science Fiction	
Mystery	
Thriller	
Key: ______ = ______ novels	

1. Explain your key and why you chose the number you did.

____

____

2. What is the total number of novels patrons borrowed on an average day?

________ novels

124 Representing Data

Use the line plot to answer the questions that follow.

The line plot below shows the heights of students in Ms. Iverson's class.

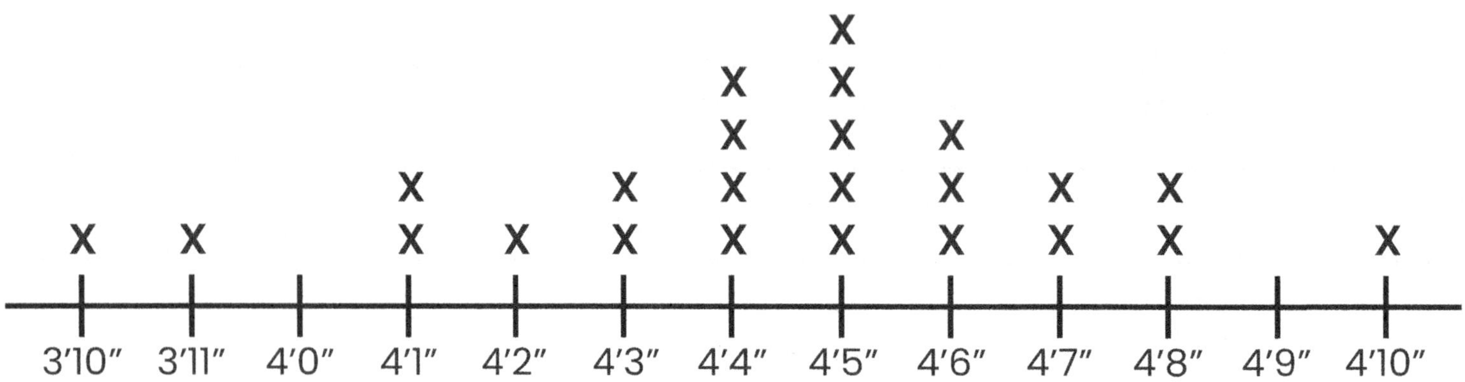

Heights of Students

1. What is the most common height in the class?

2. How many students in the class are taller than 4'0"?

3. What is the difference is height between the tallest student in the class and the shortest?

Make a line plot to show the data given. Then, answer the questions below.

Students in Mr. Hammond's class grew bean plants. The students measured the heights of their plants to the nearest centimeter. The results are shown in the chart below.

Bean Plant Height	Number of Plants
20 cm	2
21 cm	4
22 cm	4
23 cm	5
24 cm	3
25 cm	1

Create a line plot to show this data.

1. Which height was the most common?

2. How many plants had a height less than 22 cm?

3. How many bean plants were there in all?

Use the bar graph to answer the questions that follow.

Joey surveys his classmates about their favorite sport. The results are shown in the bar graph below.

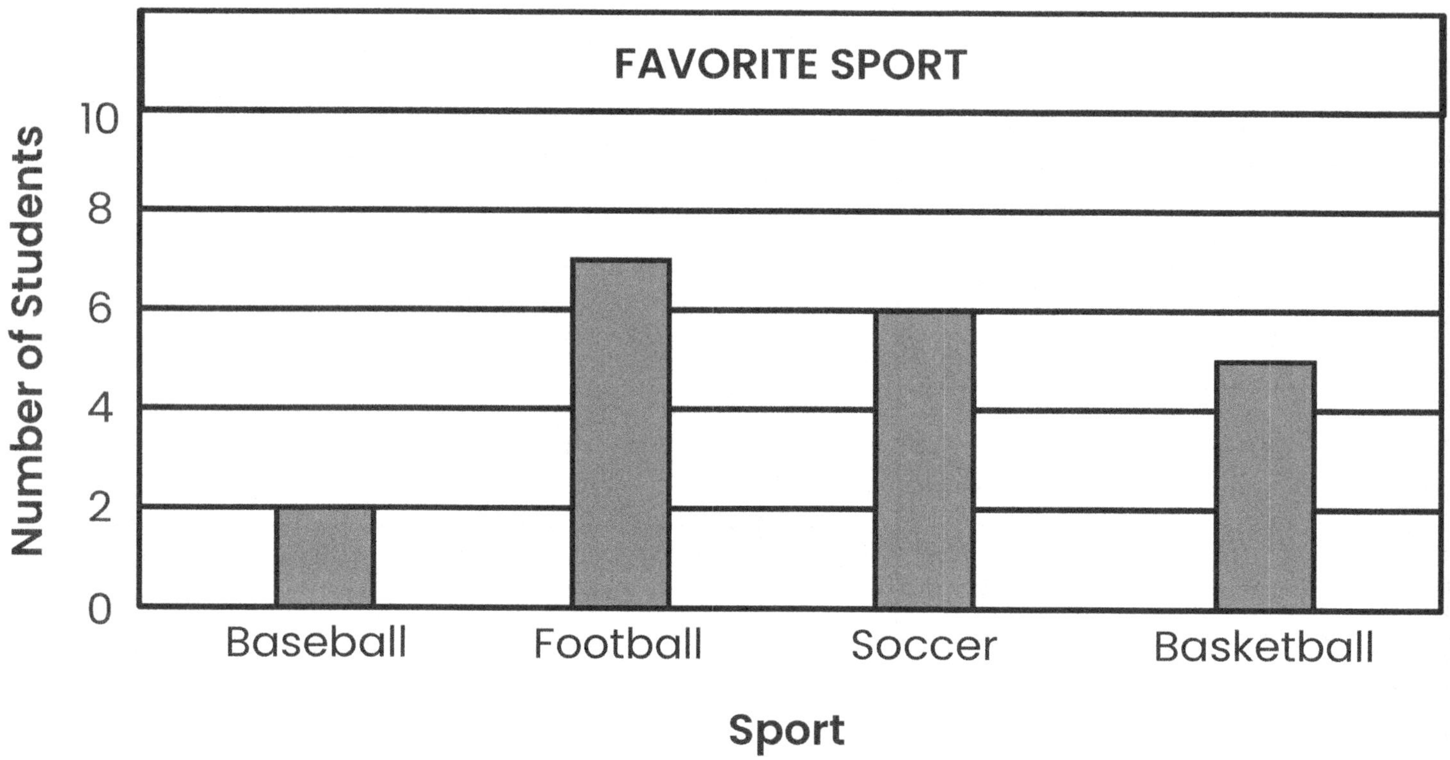

1. How many students chose football or soccer as their favorite sport?

2. How many more students chose basketball than baseball?

3. How many students did NOT choose football as their favorite sport?

Make a bar graph to show the data given.

Jenny recorded the sales of each type of food at a bake sale.

- 24 brownies
- 27 cookies
- 15 pies
- 9 cakes

Choose an appropriate scale and record the data on the bar graph below.

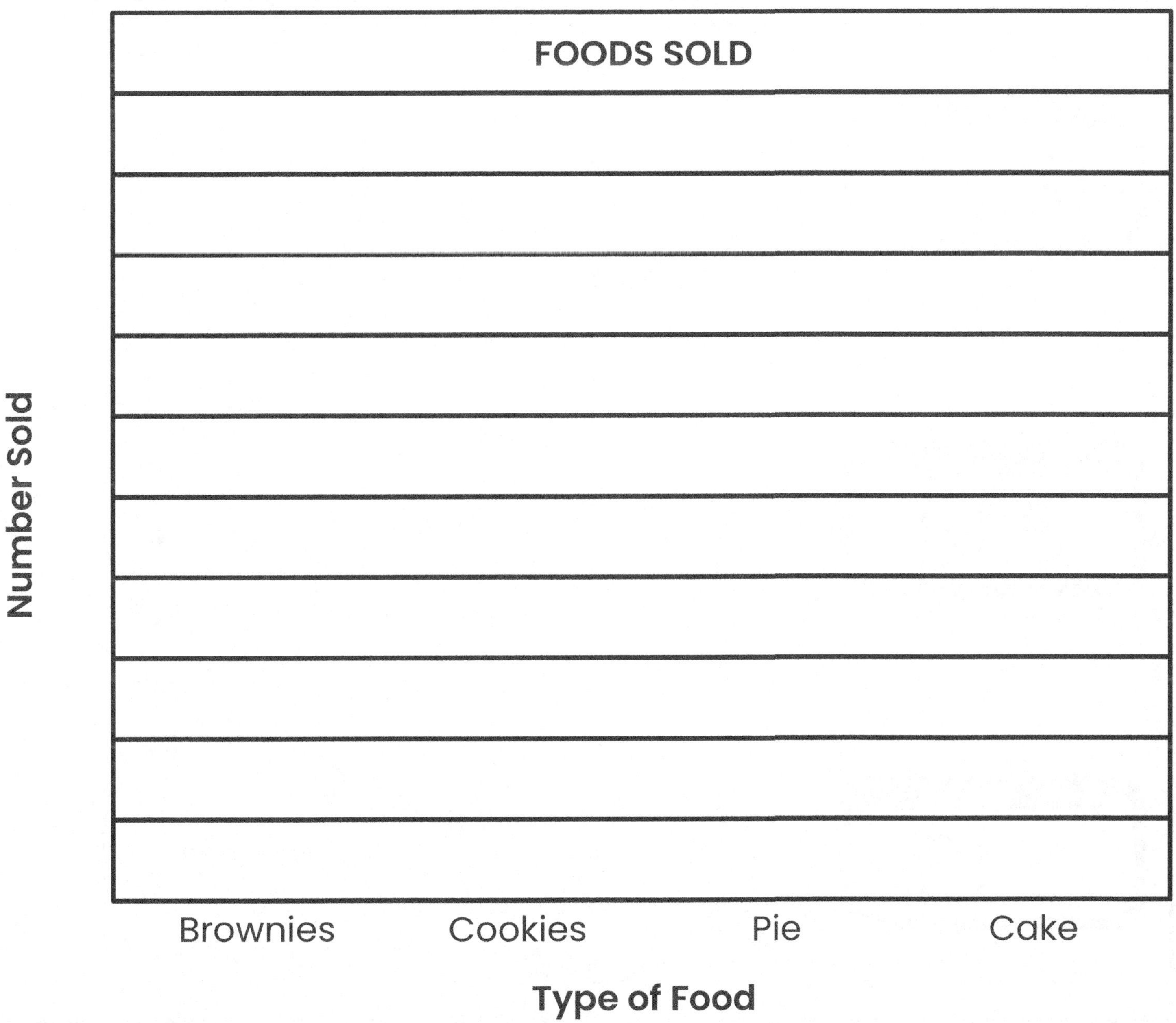

Time

Circle the analog clock in each row that shows the same time as the digital clock.

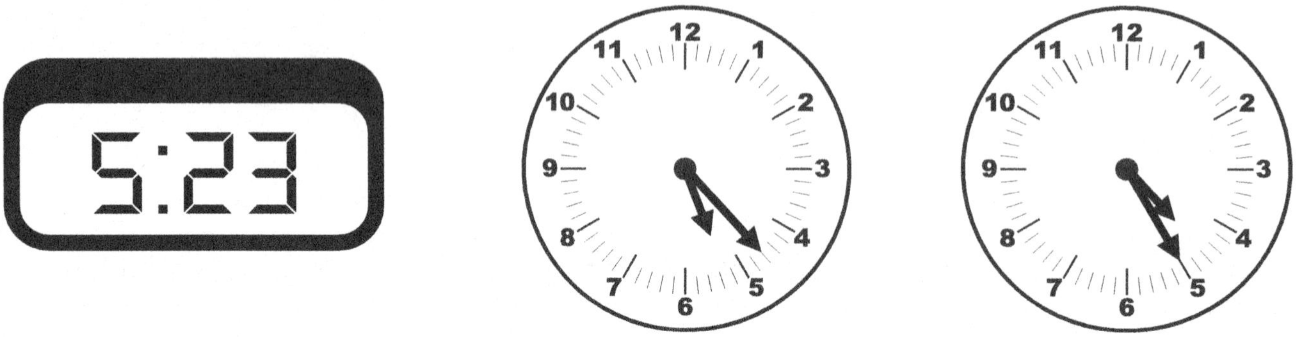

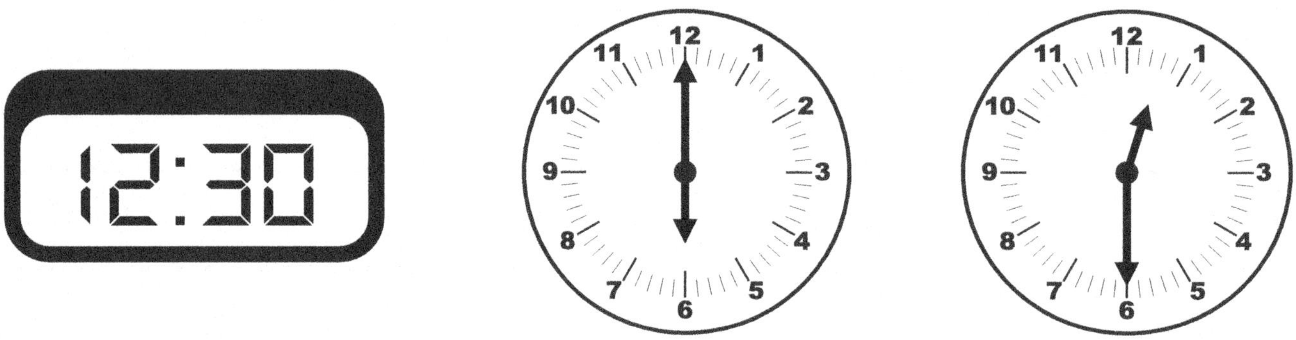

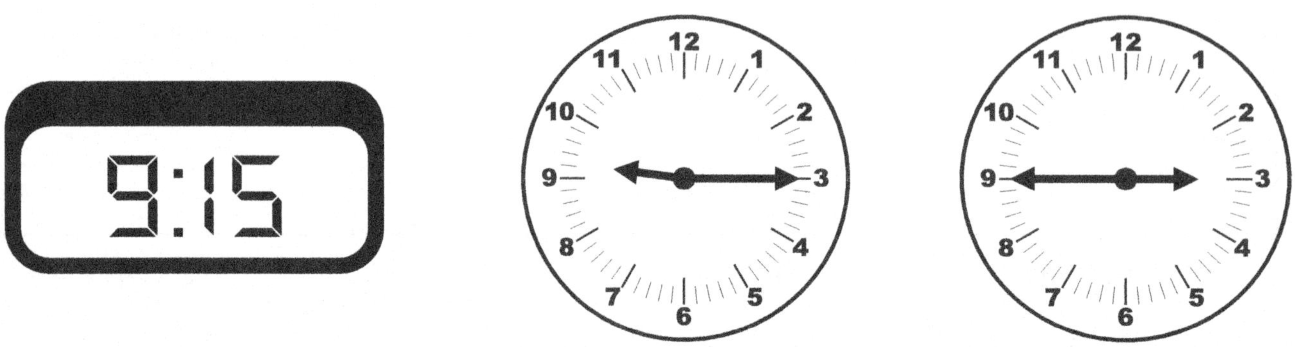

Write the time shown on each clock.

______

______

______

______

______

______

______

______

Time

Draw hands onto each clock to show the correct time.

10:17

5:42

1:30

11:03

8:35

9:54

6:19

12:25

Fill in the correct bubble to answer each question.

Which time is 35 minutes after 11:23 AM?

○ 10:48 AM ○ 11:48 AM ○ 11:58 AM

Which time is 18 minutes before 10:30 PM?

○ 10:12 PM ○ 10:18 PM ○ 10:48 PM

Which time is 20 minutes after 9:57 AM?

○ 9:37 AM ○ 10:17 AM ○ 10:27 AM

Which time is 30 minutes before 5:14 PM?

○ 4:34 PM ○ 4:44 PM ○ 5:44 PM

Which time is 1 hour and 10 minutes after 8:52 PM?

○ 9:02 PM ○ 9:52 PM ○ 10:02 PM

132 Time

Each number line below represents one hour on a clock. Write the time shown by the dot on each number line.

① ________

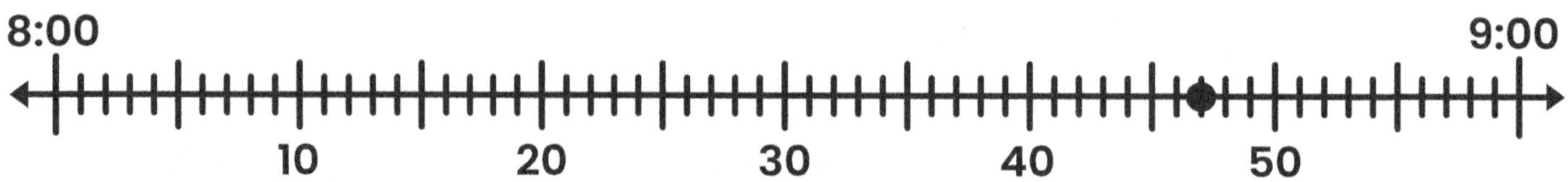

② ________

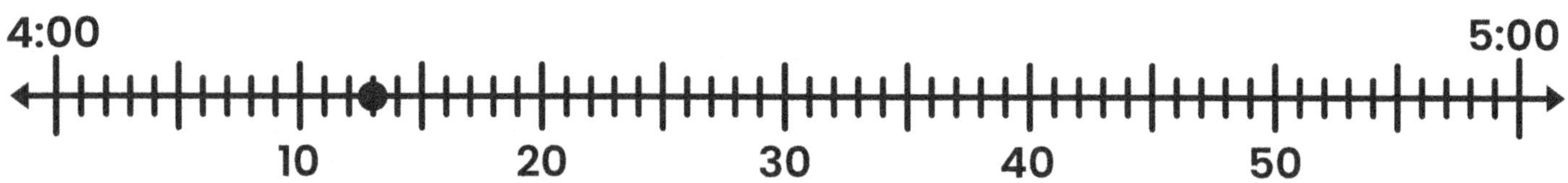

③ ________

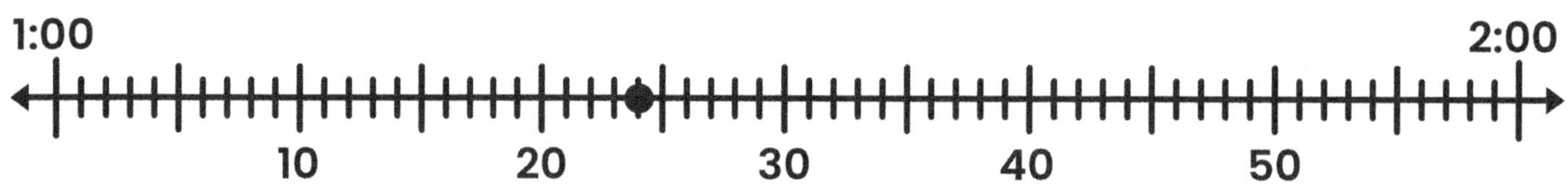

Draw a dot on the number line to show each time.

① 6:52

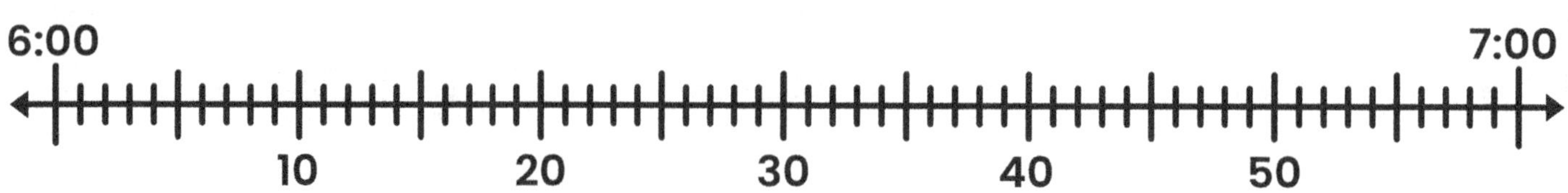

② 3:29

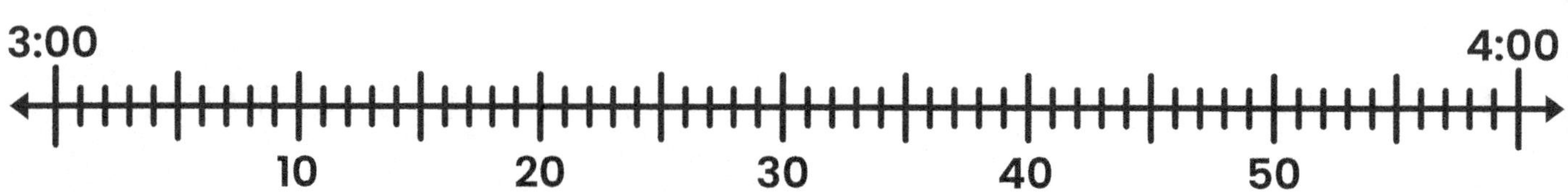

Write how many minutes have passed from point A to point B.

① ________ minutes

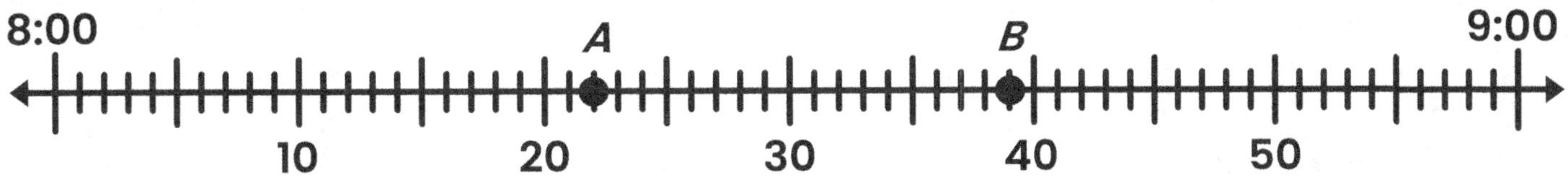

② ________ minutes

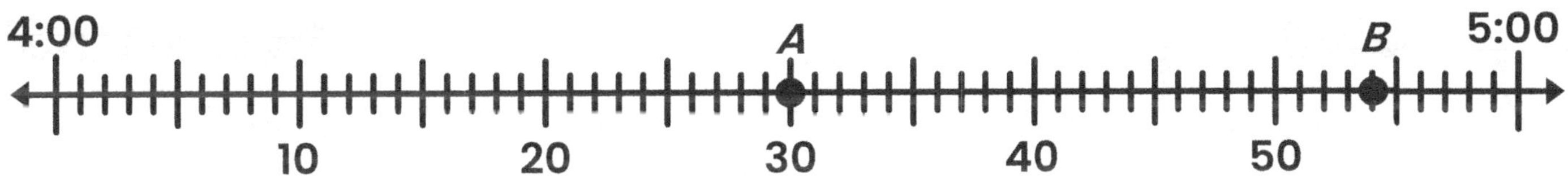

③ ________ minutes

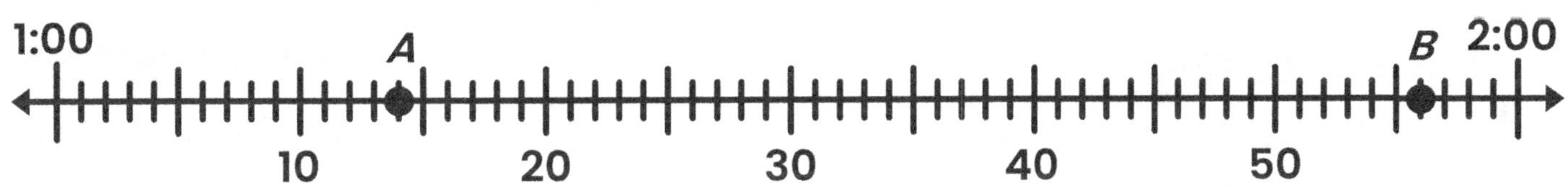

Draw points A and B on the number lines to show each time.

① Point A: 10:15 | Point B: 16 minutes later

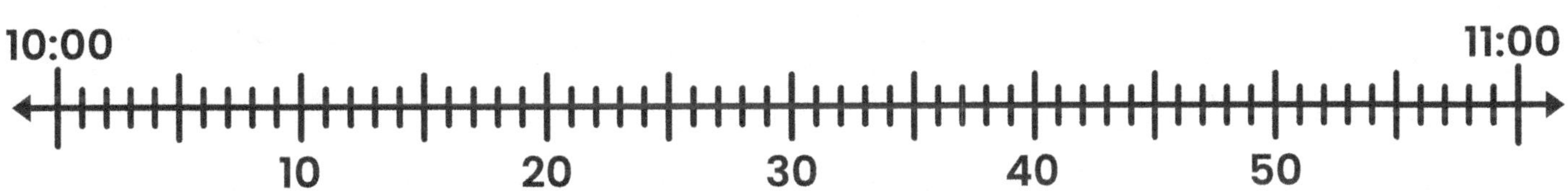

② Point A: 4:08 | Point B: 41 minutes later

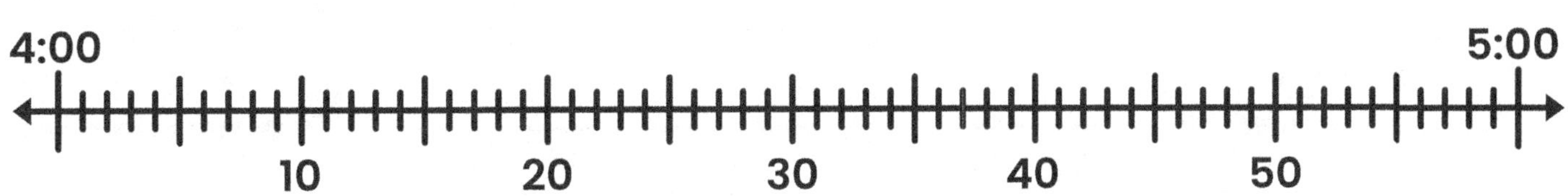

Time

Write how much time has elapsed between the time on each pair of clocks.

Start	Elapsed time	End
2:13 PM	1 hour, 23 minutes	3:46 PM
1:30 PM	____________	2:37 PM
7:24 AM	____________	7:53 AM
8:53 AM	____________	9:06 AM
5:10 PM	____________	7:15 PM

Answer each word problem about time.

The clock on the left shows the current time. Alyssa's alarm will go off at 7:00 AM. How many minutes does she have left until her alarm goes off?

The clock on the left shows the time Shawn's math class starts. The class lasts 40 minutes. What time will it end?

The clock on the left shows the time Jordan got home after riding the bus from school. If Jordan's bus left school at 3:11 PM, how long was he on the bus?

The clock on the left shows Julia's bedtime. If it is 7:05 PM now, how long does she have left until bedtime?

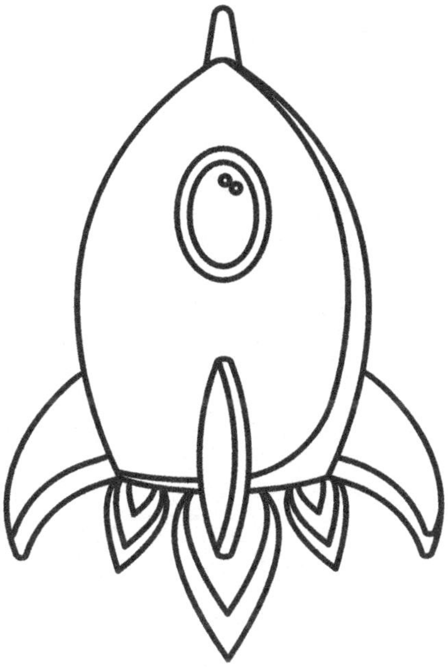

Geometry

Classifying Shapes

Circle the shapes that are polygons.

Name each shape. Write the number of sides and angles it has.

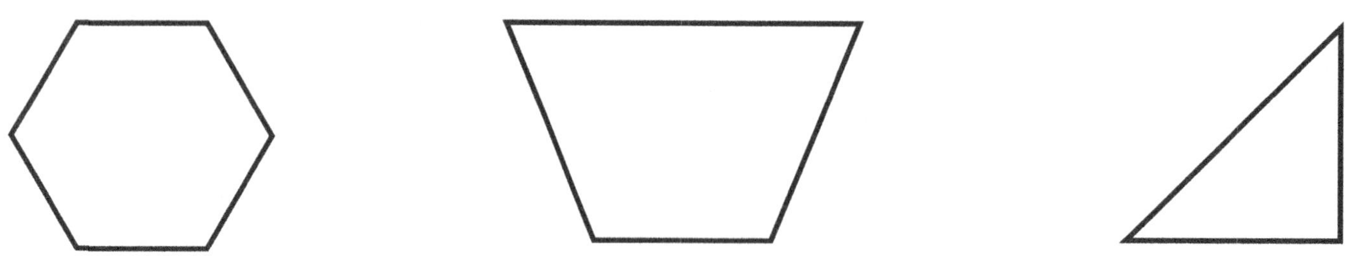

____________	____________	____________
______ sides	______ sides	______ sides
______ angles	______ angles	______ angles

Draw a small square to mark any angle that appears to be a right angle.
(Not all shapes will have one, and some may have more than one.)

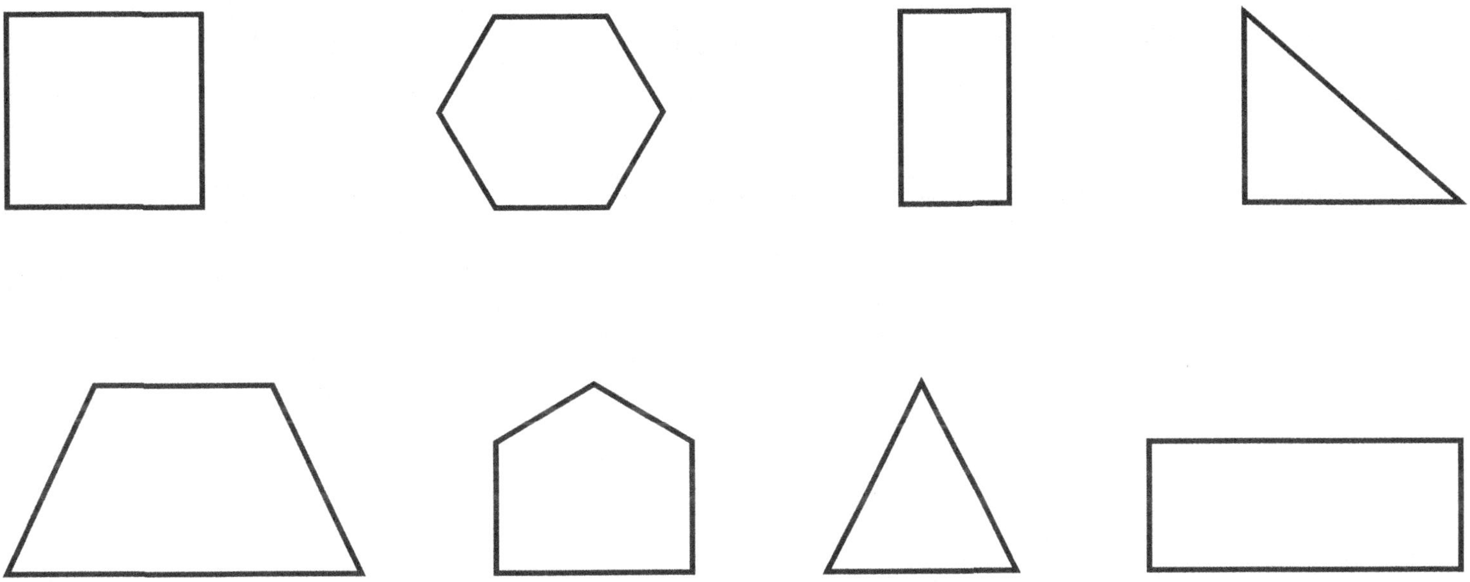

Write how many of each type of angle each shape appears to have.

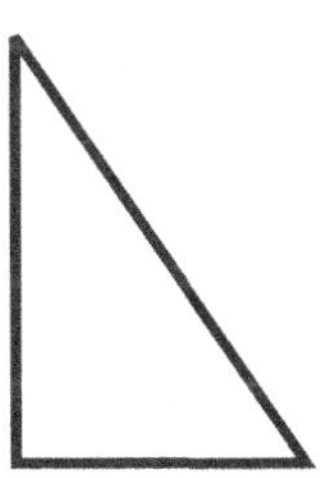

_______ Less than 90°

_______ 90°

_______ More than 90°

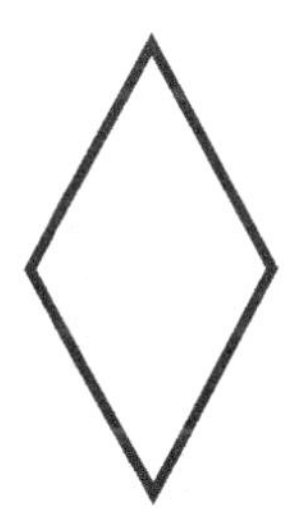

_______ Less than 90°

_______ 90°

_______ More than 90°

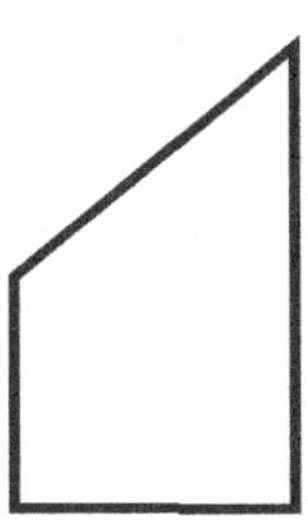

_______ Less than 90°

_______ 90°

_______ More than 90°

_______ Less than 90°

_______ 90°

_______ More than 90°

Classifying Shapes

On each shape, two sides are shown in gray. Circle whether each pair of gray sides is parallel or not parallel and equal or unequal in length.

Shape		
	Parallel Not parallel	Equal length Unequal length
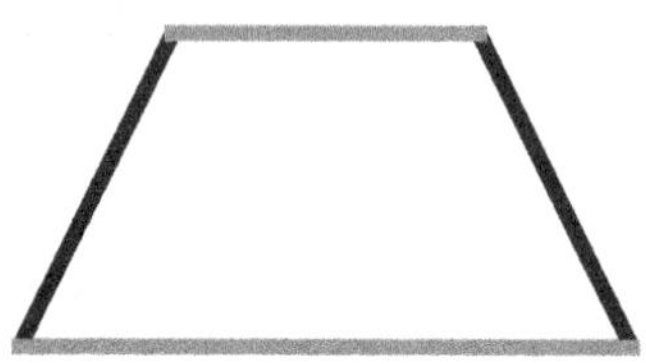	Parallel Not parallel	Equal length Unequal length
	Parallel Not parallel	Equal length Unequal length
	Parallel Not parallel	Equal length Unequal length

Classifying Shapes

Circle ALL the words that describe each shape.

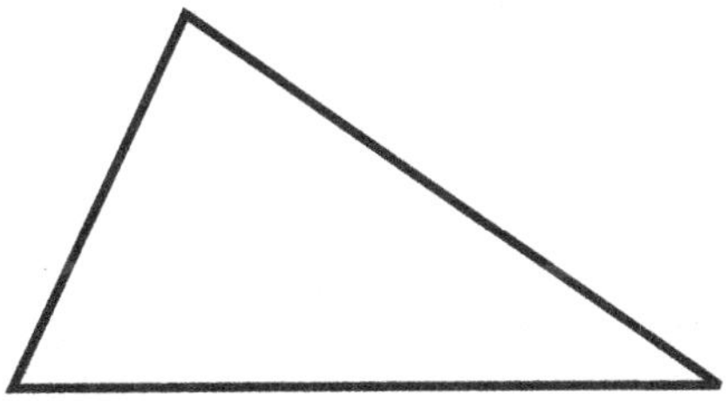

Polygon Triangle

Quadrilateral Trapezoid

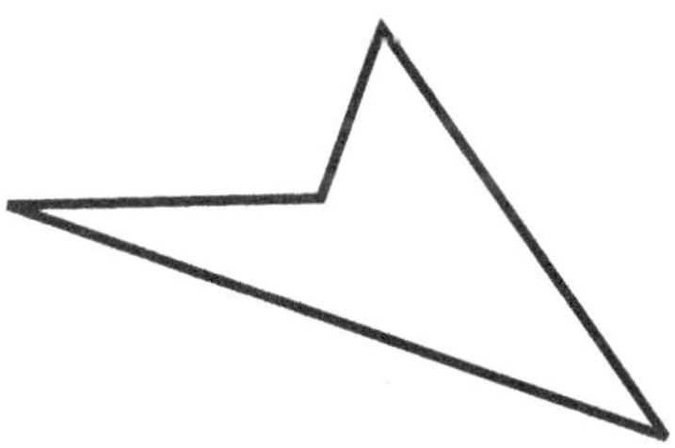

Polygon Triangle

Quadrilateral Trapezoid

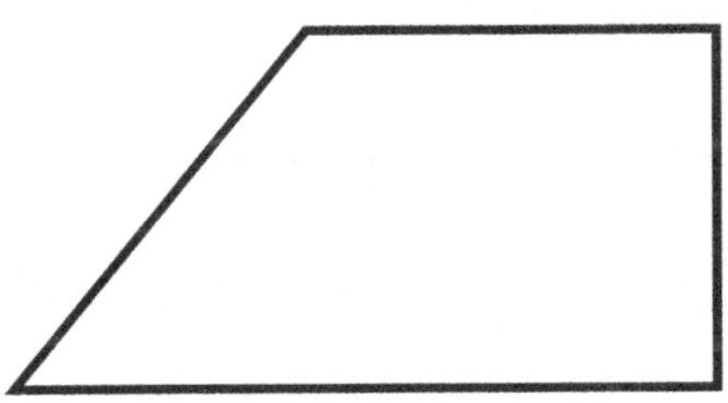

Polygon Triangle

Quadrilateral Trapezoid

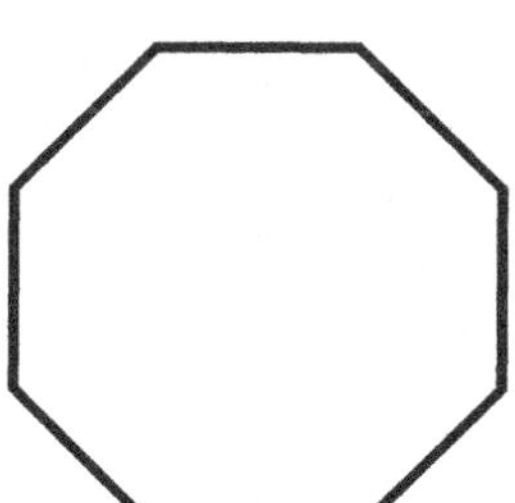

Polygon Triangle

Quadrilateral Trapezoid

Classifying Shapes

Fill in the correct bubble to answer each question about quadrilaterals.

Which quadrilateral always has 4 equal sides and 4 right angles?

○ rhombus ○ rectangle ○ square

Which of these shapes always has 2 pairs of parallel sides?

○ parallelogram ○ trapezoid ○ kite

Which of these shapes could have an angle less than 90°?

○ rectangle ○ square ○ parallelogram

Which of these could have adjacent sides that are NOT equal?

○ square ○ parallelogram ○ kite

Which of these could have an angle greater than 90°?

○ trapezoid ○ square ○ rectangle

Classifying Shapes

Circle ALL the words that describe each quadrilateral.

Parallelogram	Rhombus
Square	Rectangle

Parallelogram	Rhombus
Square	Rectangle

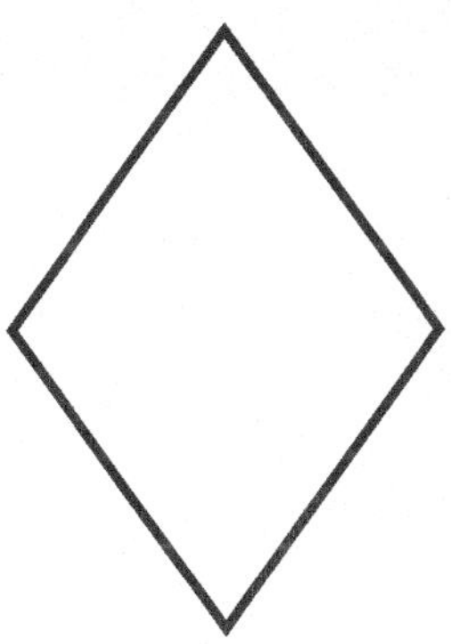

Parallelogram	Rhombus
Square	Rectangle

Parallelogram	Rhombus
Square	Rectangle

144 Perimeter

Find the perimeter by counting on the grid.

① What is the perimeter of the shape shown on the grid?

_______ units

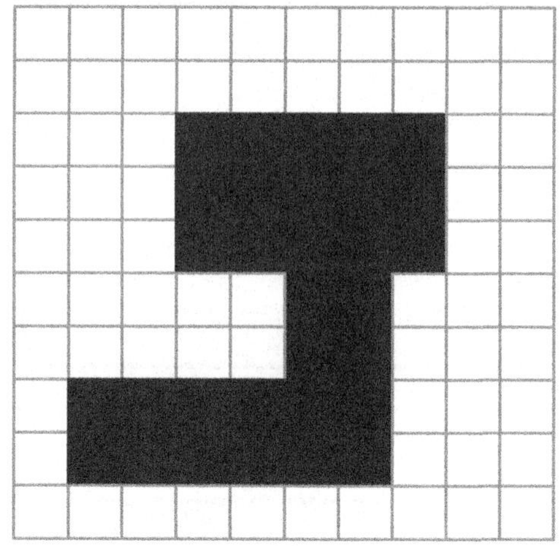

② What is the perimeter of the shape shown on the grid?

_______ units

Draw a shape on each grid that has the perimeter shown.

③ 20 units

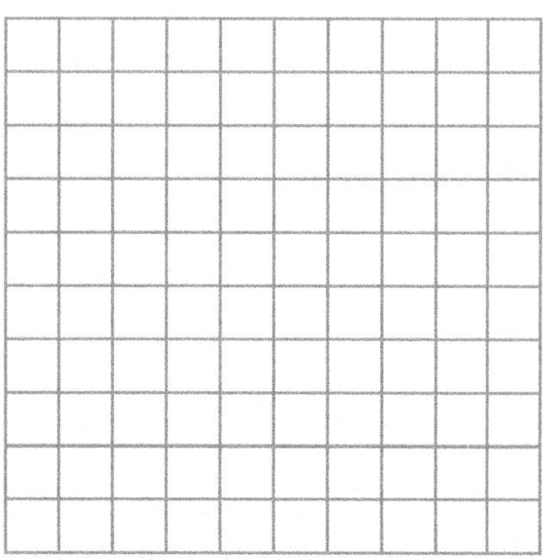

④ 16 units

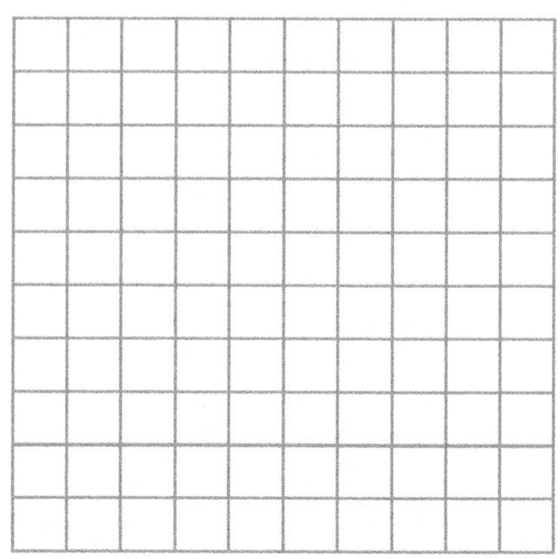

Perimeter

145

Find the perimeter of each shape.

1

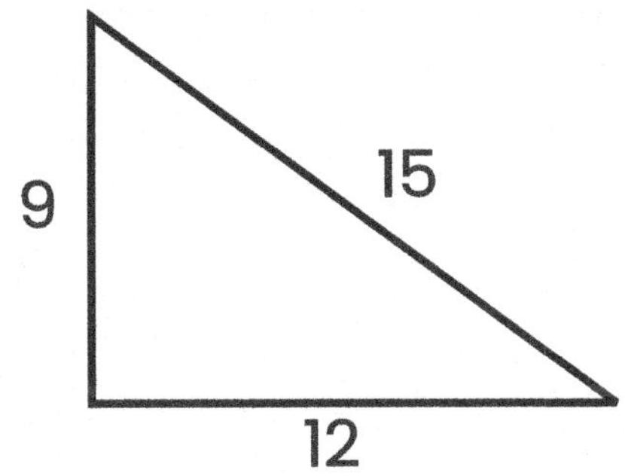

_______ units

2

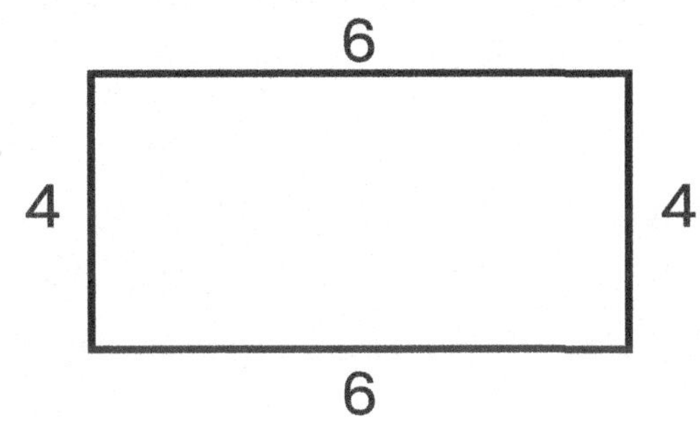

_______ units

3

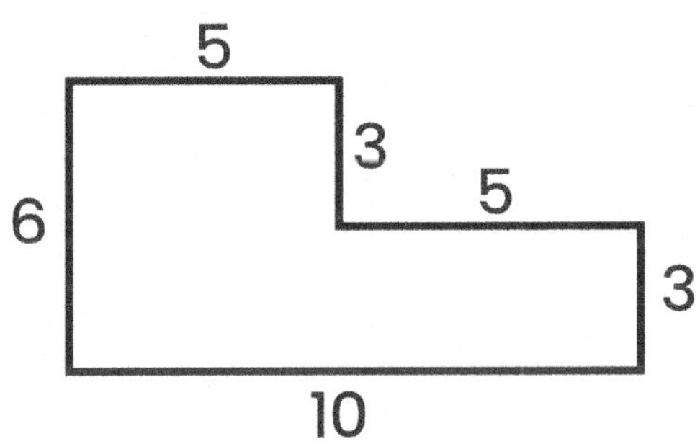

_______ units

4

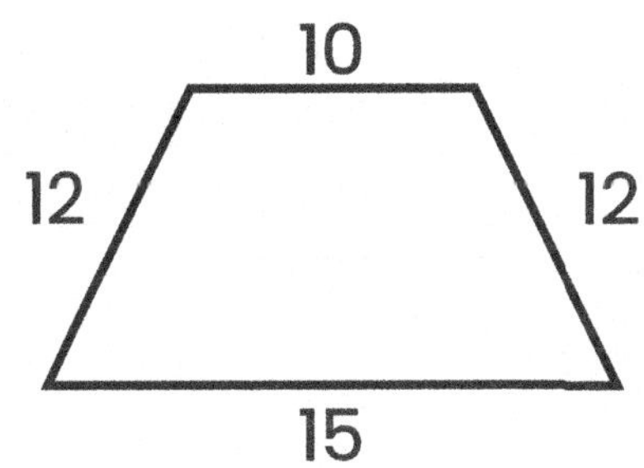

_______ units

5

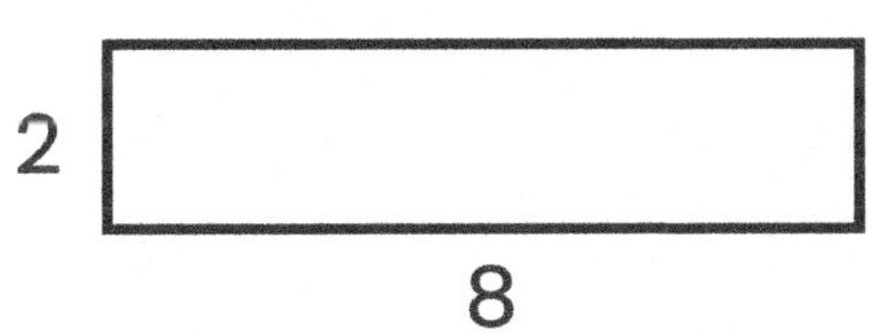

_______ units

6

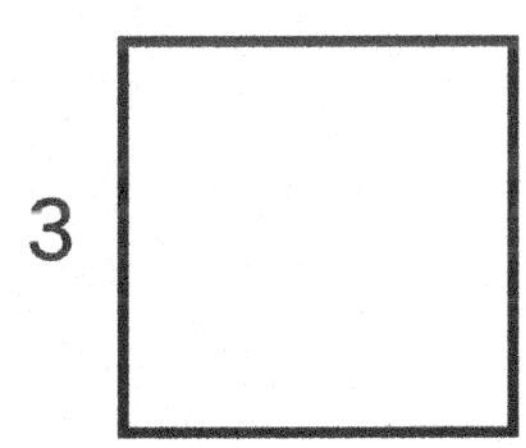

_______ units

Perimeter

Answer each word problem about perimeter.

A picture frame measures 18 inches by 20 inches. What is the perimeter of the frame?

________ inches

Betty is building a fence around her rectangular garden, which measures 12 feet by 15 feet. What will be the total length of the fence?

________ feet

A rectangular piece of paper measures 21 cm by 29 cm. What is the perimeter of the piece of paper?

________ cm

A sports field measures 60 meters by 100 meters. What is its perimeter?

_________ meters

Melody has a square rug with a perimeter of 32 feet. How long is each side of the rug?

_________ feet

Carl is putting baseboard trim around the edges of a 10 foot by 12 foot bedroom. The trim costs $1 per foot. What will be the total cost of the trim for the room?

____

148 Area

Each shape is made up of unit squares. Count to find the area of each shape in square units.

1

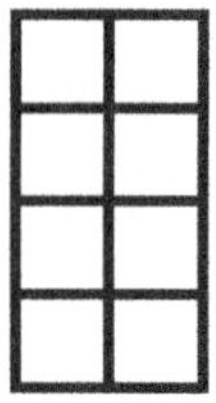

______ square units

2

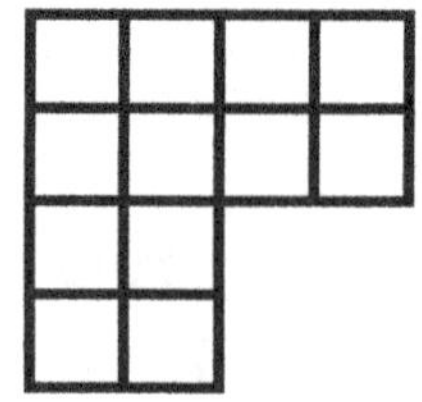

______ square units

3

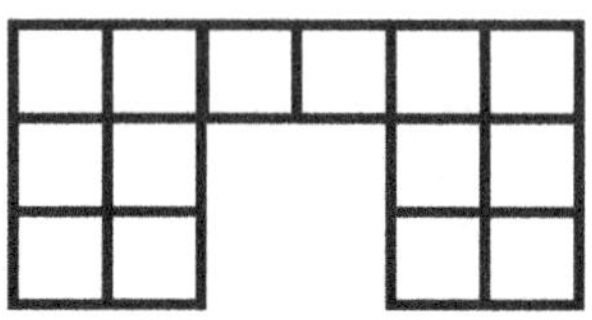

______ square units

4

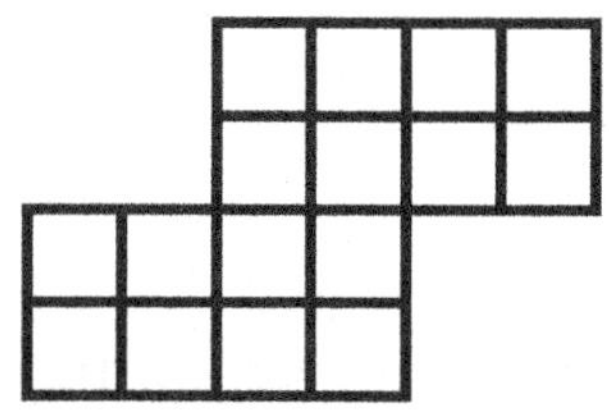

______ square units

Draw a shape on each grid that has the area shown.

5 12 square units

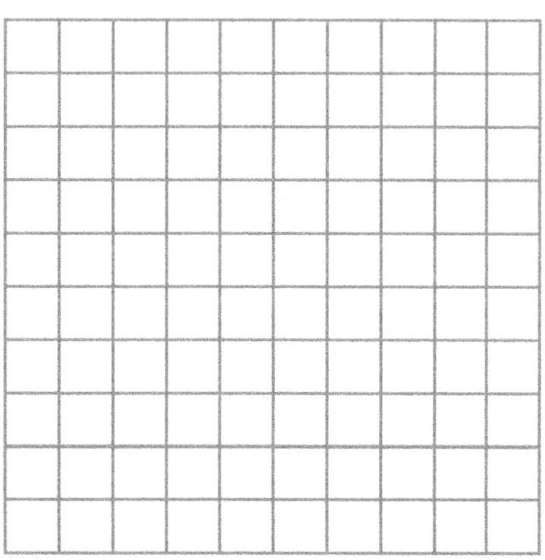

6 20 square units

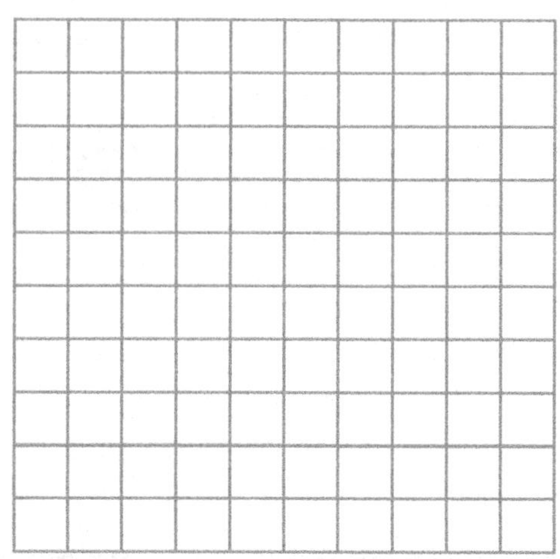

Color the shapes in each row that have the same area as the first shape.

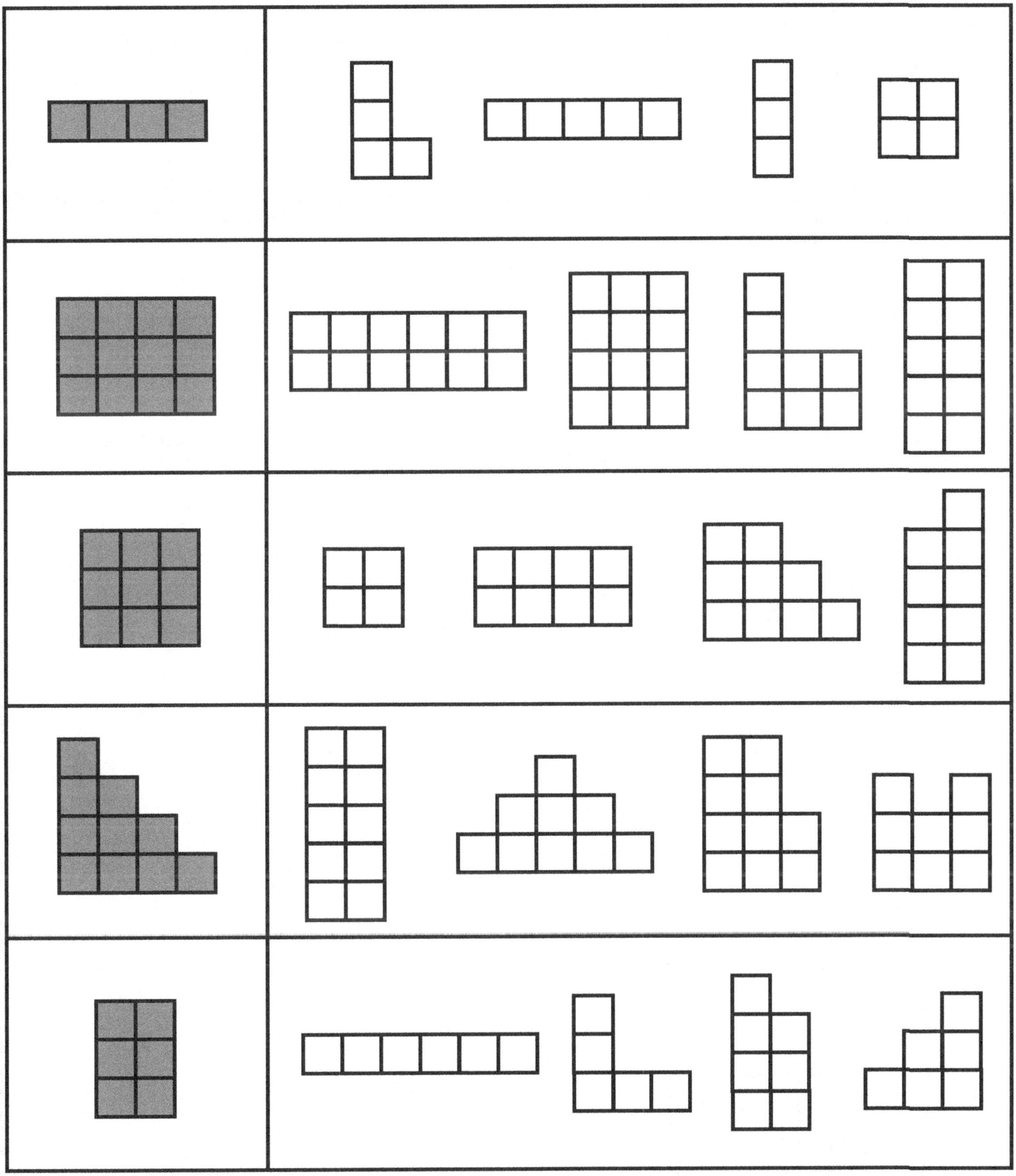

150 Area

Each rectangle is made of unit squares. Multiply the length by the width to find the area.

(EX) 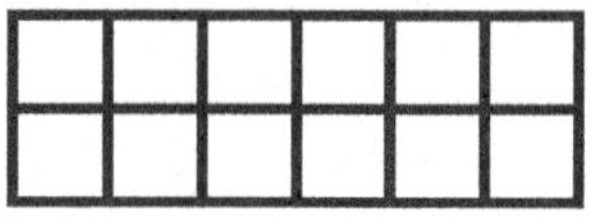

__6__ × __2__ = __12__ square units

(1) 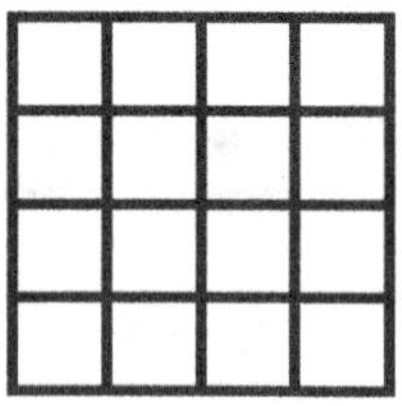

______ × ______ = ______ square units

(2) 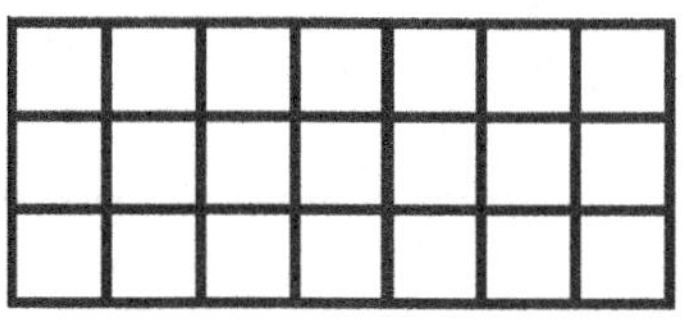

______ × ______ = ______ square units

(3) 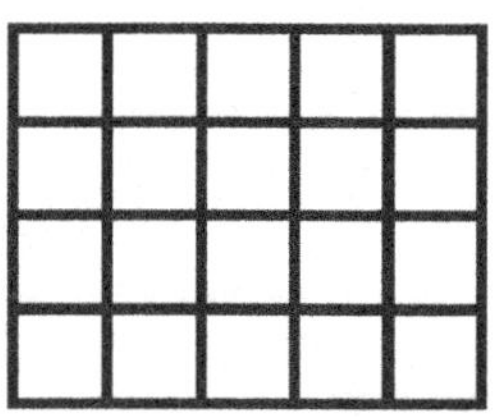

______ × ______ = ______ square units

(4) 

______ × ______ = ______ square units

(5) 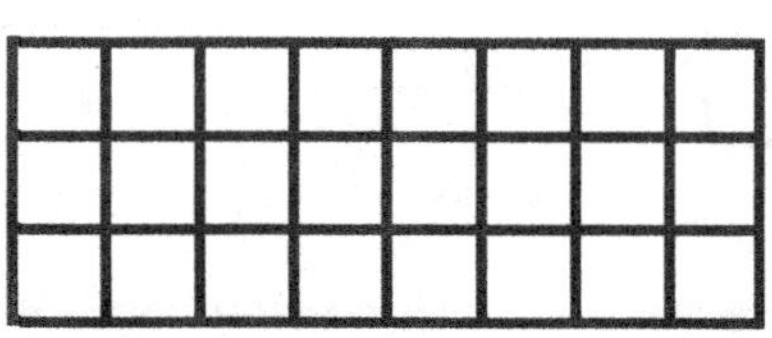

______ × ______ = ______ square units

Multiply the length by the width to find the area of each rectangle.

1.

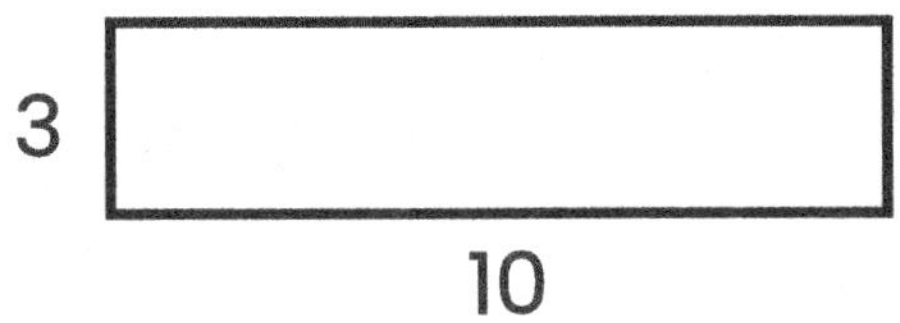

_______ square units

2.

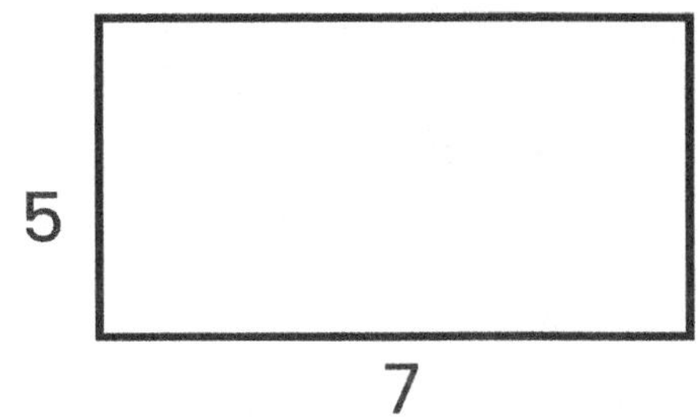

_______ square units

3.

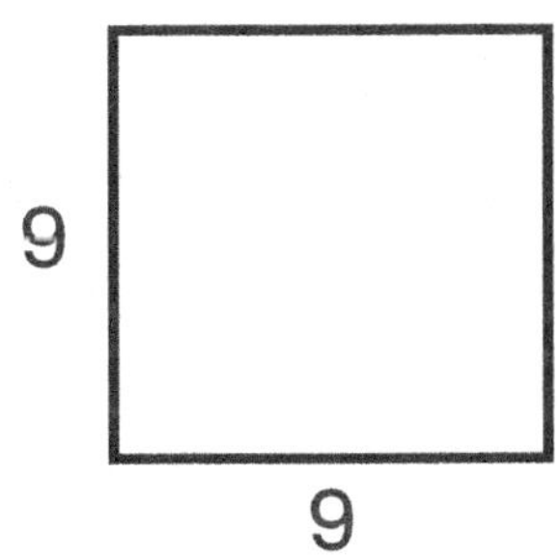

_______ square units

4.

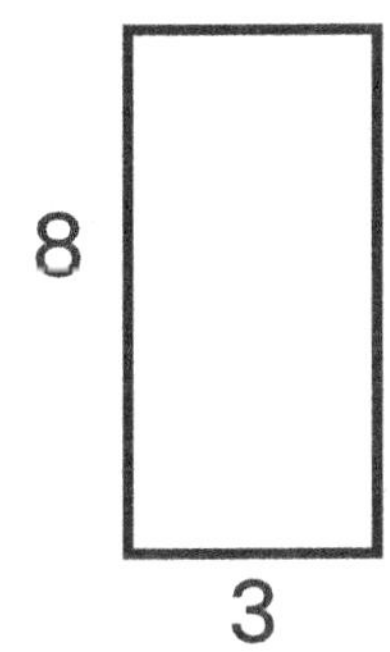

_______ square units

5.

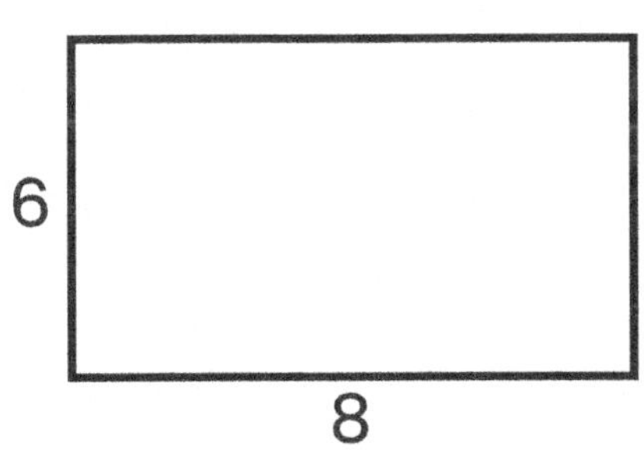

_______ square units

6.

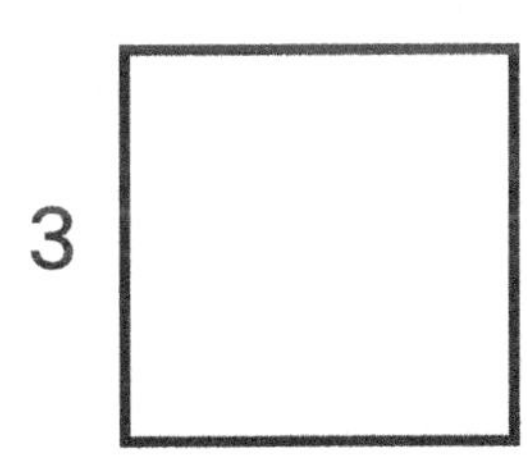

_______ square units

Area

Use the area and divide to find the missing side length of each rectangle.

EX

	Area = 24 square units
?	
	8

24 ÷ 8 = 3 units

1

	Area = 28 square units
?	
	7

_____ ÷ _____ = _____ units

2

	Area = 45 square units
5	
	?

_____ ÷ _____ = _____ units

3

	Area = 48 square units
6	
	?

_____ ÷ _____ = _____ units

4

	Area = 18 square units
?	
	6

_____ ÷ _____ = _____ units

5

	Area = 24 square units
4	
	?

_____ ÷ _____ = _____ units

Solve each word problem involving area.

1. Rachel is buying new carpet for her living room. The rectangular room measures 5 meters by 6 meters. How many square meters of carpet will she need?

   ______ sq. meters

2. The deck in Malcolm's backyard measures 8 feet by 10 feet. What is the area of the deck?

   ______ sq. feet

3. Everett is tiling his bathroom floor with square tiles that measure 1 foot by 1 foot. The bathroom measures 5 feet by 8 feet. How many tiles will he need to cover the floor?

   ______ tiles

4. Gabby hangs two posters on her wall. Each one measures 2 feet by 3 feet. What is the total area of the two posters?

   ______ sq. feet

5. A rectangular sign has an area of 63 square feet. If the width of the sign is 7 feet, what is the length of the sign?

   ______ feet

154 Area

Draw a line to divide each shape into two rectangles. Find the area of each, then add to find the total area.

EX

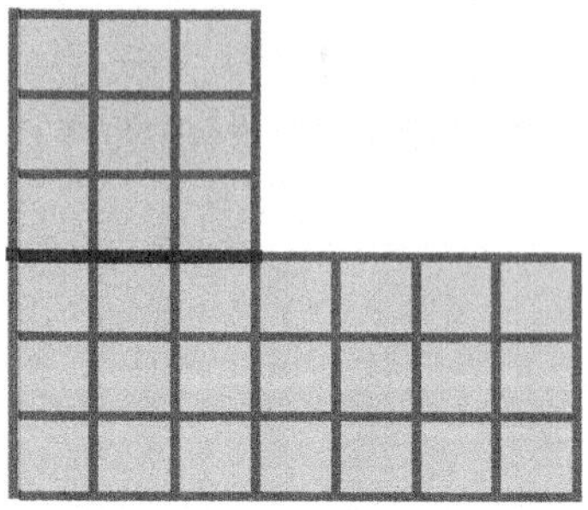

Area 1 = 9 square units

Area 2 = 21 square units

9 + 21 = 30 square units

1

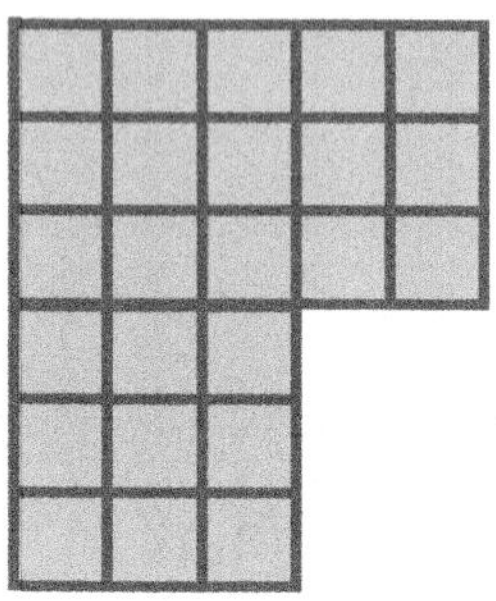

Area 1 = ______ square units

Area 2 = ______ square units

______ + ______ = ______ square units

2

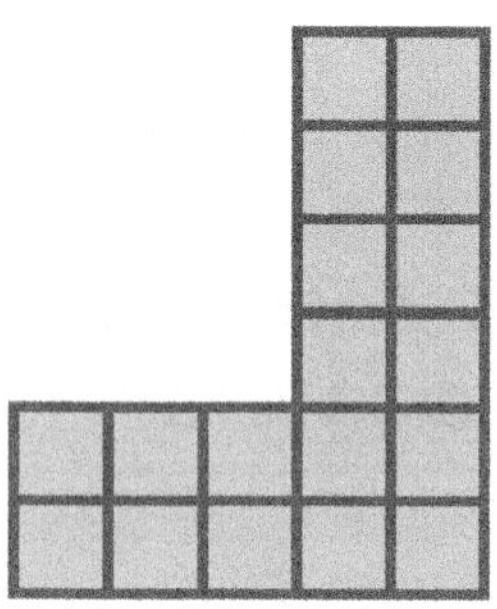

Area 1 = ______ square units

Area 2 = ______ square units

______ + ______ = ______ square units

3

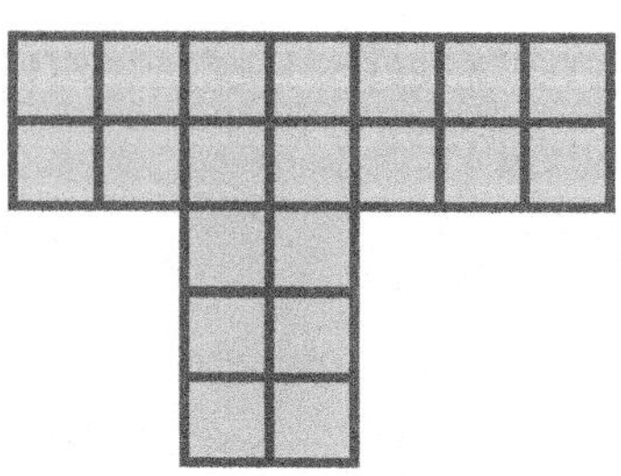

Area 1 = ______ square units

Area 2 = ______ square units

______ + ______ = ______ square units

Draw a line to divide each shape into two rectangles. Find the area of each, then add to find the total area.

(EX)

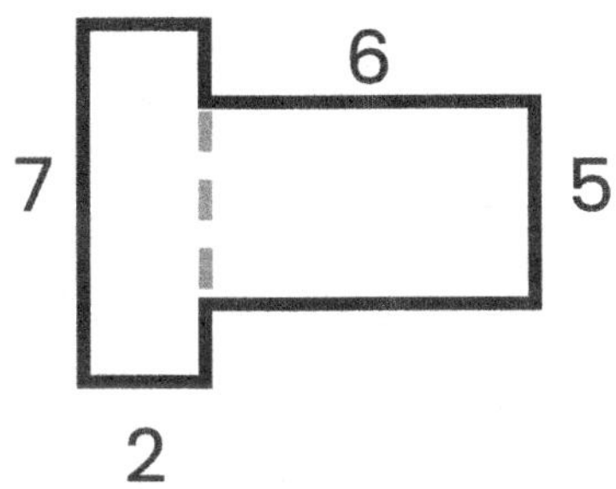

Area 1 = 14 square units

Area 2 = 30 square units

14 + 30 = 44 square units

(1)

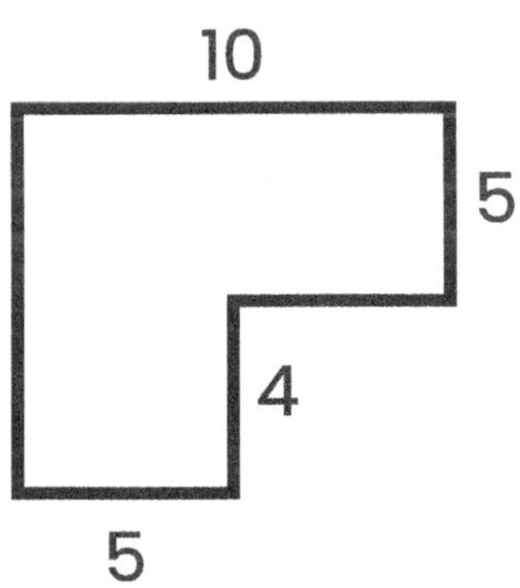

Area 1 = ______ square units

Area 2 = ______ square units

______ + ______ = ______ square units

(2)

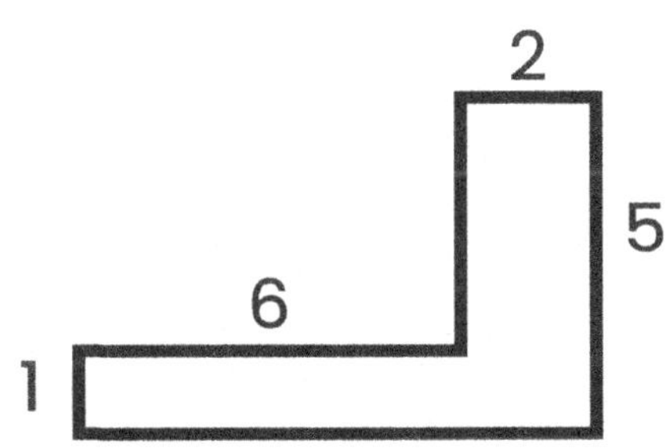

Area 1 = ______ square units

Area 2 = ______ square units

______ + ______ = ______ square units

(3)

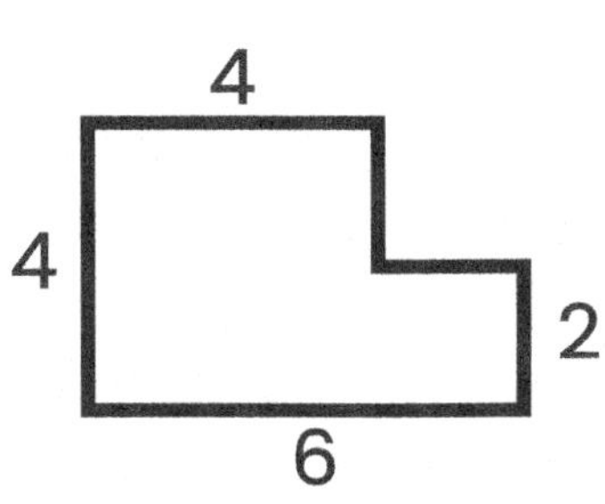

Area 1 = ______ square units

Area 2 = ______ square units

______ + ______ = ______ square units

Made in the USA
Las Vegas, NV
27 March 2024

87819919R00085